THE STUDENT ENVIRONMENTAL ACTION GUIDE

THE STUDENT ENVIRONMENTAL ACTION COALITION

Dedicated to students—past, present, and future—who believe that changing their campuses is the first step in changing the world.

Copyright © 1991 by Javnarama / The EarthWorks Group.
All Rights Reserved.
Printed in the United States of America.
No part of this book may be used or reproduced in
any manner whatsoever without written
permission, except in the case of brief quotations
embodied in critical articles or reviews.

Created and packaged by Javnarama
Designed by Javnarama

ISBN 0-06-500432-9
First Edition 10 9 8 7 6 5 4 3 2 1

We've provided a great deal of information about
practices and products in this book. In most cases,
we've relied on advice, recommendations and research by
others whose judgments we consider accurate and free
from bias. However, we can't and don't guarantee
the results. This book offers you a start. The
responsibility for using it ultimately rests with you.

Printed On Recycled Paper

ACKNOWLEDGMENTS

SEAC would like to thank everyone who worked with us to make this book possible, including:

- John Javna
- Chris Fox
- Di Goodman
- Lyn Speakman
- Fritz Springmeyer
- Phil Catalfo
- April Smith
- Renee Robinson
- Craig Reisser
- Dianell Dreibelbis
- Beth Ising
- Cris Moore
- Will Toor
- Nick Keller
- Kristen Brown
- Yewande Dada
- Julian Keniry
- National Wildlife Federation
- *Cool It!*
- Catherine Dee
- Melanie Foster
- Dayna Macy
- Lenna Lebovich
- John Dollison
- Megan Anderson
- Lynn Davies
- Jay Nitschke
- Sharilyn Hovind
- Earth Day 1990
- Helen Denham
- J Burger
- Eric Odell
- Ericka Kurz
- Randy Viscio
- Andy MacDonald
- Lee Altenberg
- Public Interest Research Groups
- Flora Lu

- Miya Yoshitani
- Brian Trelstad
- Erin Goewey
- Spencer Weisebroth
- Jennifer Dill
- David Kupfer
- Tom DeVries
- YouthAction
- Partnership for Democracy
- David Thorne
- Gary Cerkvenik
- Owen Byrd
- Morgan Grove
- Howard Brotine
- The Joy of Garbage
- Christian Becker
- Christine Larson
- Willard Dakin
- Eric Kessler
- Jeanette Galanis
- Lara Mears
- Ben Pargman
- Rob Mahedy
- Mud Baron
- Allison Cunfer
- Ron Dinchak
- Karen Plaut
- Eric Meyer
- Adam Berrey
- Dave Hodges
- Taylor Root
- Kirti Shastri
- Jason Halbert
- Leon Brown
- Howard Carbone
- Julie Blackburn
- Paula Simpson-Grutzmacher
- Adam Walker

- Joi Elizabeth Nis
- Kristen D'Atri
- UCLA Audit Team
- Tom Swan
- Karen Stults
- Dan Damelin
- Alec Guettel
- Preston Holmes
- Jennifer Schoen
- Jennifer Karson
- Chris Kelley
- John'O Niles
- Elizabeth Wheat
- United States Student Association
- GROW Training
- Laura Fitton
- Yale Env. Center
- Rachel Fox
- Jim Williams, NACS
- Steve Kemler
- James Corless
- Shana Morrow
- Anne Steffy
- Marty Chintala
- Phillipe Hensel
- Joan McKearnan
- Mike Guilfoyle
- Tom Elliott
- Nate Boone
- Jenny Sowerine
- Sabrina Alvarz
- Justin Lehner
- Kimberly Kluff
- Michelle Anthony
- Tracy McCleod
- Mike Martin
- Tom Elliott
- Ruth Hunter
- And many more

CONTENTS

Introduction 6

1. Register to Vote..9
2. Mug Shot 12
3. Too Much Food 15
4. Everyone Into the Pool............................18
5. Cut Paper 21
6. Light Right 24
7. Tree's Company......................................28
8. Environmental Internships.....................32
9. The Right Path.......................................35
10. Scale It Down..39
11. Don't Can It...42
12. Hold an Eco-lympics 45
13. A Rottin' Thing to Do 48
14. Show Some Class 52
15. Green Your Bookstore...........................55
16. Water You Doing?.................................58
17. Go Public 62
18. Promote Energy Conservation 65
19. Start a Recycling Program....................69
20. Bug Off 73
21. Precycle 77
22. Branch Out...81
23. Buy Recycled 84
24. What's for Dinner?................................87
25. Build a Green House 90

Stay Involved..93
Join SEAC 96

A Note from the Earthworks Group

Our book *50 Simple Things You Can Do to Save the Earth* was among 1990's top bestsellers in college bookstores.

This told us that students are eager to know more about protecting our fragile environment.

But while students could *use 50 Simple Things*, it wasn't specifically designed to address campus environmental issues.

This book is. It's written for students...by students.

The Student Environmental Action Coalition has put together an impressive combination of facts, success stories, and prescriptions for action. We're proud to have been a part of it.

Traditionally, students have been a powerful force for creating social change. And today, no issue is more important than changing the way we treat the Earth.

We believe that, using their extraordinary energy and vision, students can make this change possible.

Everything depends on it.

INTRODUCTION

When we started the Student Environmental Action Coalition in 1989, we hoped that a few hundred students might find it helpful.

So we organized Threshold, the first national student environmental conference. It drew 1,700 students to Chapel Hill, North Carolina. The next year, in 1990, we organized Catalyst, the second SEAC national conference. This time we brought 7,600 students together in Champaign, Illinois. It was the largest event of its kind in history.

In two years, SEAC has grown to become the largest student-run, student-led environmental organization in the nation, with members at over 1,500 colleges and high schools in all 50 states. There are now chapters in 16 countries as well.

Today, SEAC is:
• Broadening the base of student support for the environment by organizing across lines of class, race, gender, nation, geography, and issue.
• Working to encourage our nation's corporations to live up to high environmental standards through the "Corporate Accountability Campaign."
• Acting globally to develop an international program to unite students in 86 countries.
• Organizing a campus environmental audit program to create a blueprint for environmental change on campus.

We're happy to report that we get calls every day from students. Some call with ideas and information; others call to ask what they can do to combat environmental problems. SEAC was created to answer these questions—to empower students by providing them with the tools they need to take action.

But we know that not everyone's going to call us. So we've created this book. *The Student Environmental Action Guide* is designed to show you two things:

- What you can do, either on your own or with other students
- What students are already doing to protect the planet

The projects in this book are called "simple things" because anybody can do them; you don't need to be an expert. You do, however, need a vision of what you're trying to accomplish, why it's important, and a clear sense of how to do it. That's what we've tried to give you.

This isn't a comprehensive guide—it doesn't try to tell you every detail. And because of space constraints, there are many success stories that we haven't been able to include. But the projects are carefully laid out, and there are plenty of resources available to get you and your group started.

As you become more active, we hope you will join us at SEAC. Together, we can make a difference.

The Student Environmental
Action Coalition
July 10, 1991

WHAT YOU

CAN DO

1. REGISTER TO VOTE

Your vote can make a big difference. During the 1980s in California alone, more than 21 local elections were decided by a single vote.

It's September and you just got to school. You barely registered for classes in time—and now someone's already telling you to register for something else.

Don't worry, this one's easy...but essential. As the head of the League of Conservation Voters says: "The single most important leverage we have as concerned environmentalists is to participate in the electoral process."

If you really care about protecting the Earth, make registering to vote a priority when you turn 18...and keep re-registering every time you move.

TRULY DE-VOTED

• An estimated 14 million college students make up about 8% of the potential voting population. But so few Americans participate in national elections that students could control as much as 20% of the vote.

• Unfortunately, students aren't using their power. According to U.S. census figures: Only 62% of all college students were registered to vote for the 1988 elections...and only 47.9% actually voted.

• Voting in a college town can have national impact. Many environmental policies—e.g., the original styro-ban in Berkeley, California—get their start in communities with large student populations.

• In 1960, John F. Kennedy was elected president by less than one vote per precinct. If only a handful of voters in each town hadn't bothered to vote that day, he wouldn't have been elected.

REGISTRATION BASICS

• If your college address changes, remember to update your voter registration. Otherwise, you may not be able to vote.

• If you register to vote at home, but will be away at school, apply

for an absentee ballot. Important note: check the deadlines; they vary from state to state.
- It's easy to register on college campuses at election time. Look for voter registration tables in front of the student union, cafeteria, etc.
- You can also register at post offices, libraries, motor vehicle departments, or by mail.
- In all states except North Dakota, you must register to be able to vote. In all states except Maine, Minnesota, and Wisconsin, you must register before election day. Registration deadlines vary, but in most states it's 30 days before the election. Check with your Registrar of Voters.

WHAT YOU CAN DO

On Your Own
- **Register to vote.** You can register either at your college address or in your hometown.
- **Vote green.** Call local environmental groups to find out which candidates and ballot measures they support.
- **Send for a League of Conservation Voters environmental scorecard.** It lists how all U.S. senators and members of the House of Representatives voted on key environmental bills. Cost: $5. (See Resources for address.)

With Other Students
- **Put together a local environmental scorecard.** Distribute it around your campus and community. (The League of Women Voters can help you compile candidates' voting records.)
- **Work with your campus newspaper** to publish local candidates' environmental records....Or publish your environmental scorecard as an ad; University of Wisconsin students did in 1991, and got 1,500 more students to register.
- **Maximize voter turnout.** Requisition a campus vehicle on election day and transport students to and from the polls.
- **Organize a voter registration drive.** (See Resources.)
- **Build a coalition...and run for office.** Christine Larson, an environmentalist and a student at the University of Wisconsin, did it. She ran for a seat on the Madison City Council...and won.

CAMPUS SUCCESS STORY
Stanford University, Palo Alto, California

Background
In 1990, a group of Stanford students joined forces with Bay Area Action (BAA), a local environmental group, to get out the "green" vote on campus. They used BAA's office as home base.

What They Did
1. With the help of groups like the California Public Interest Research Group, they obtained lists of students who'd voted for environmental legislation in the past. Their plan: Contact all of them.
2. To make canvassing easier, they divided the campus into precincts (250-300 people in each) and appointed precinct leaders.
3. Each precinct leader recruited volunteers to help with the work.
4. First, they made phone calls telling students about important environmental initiatives in the upcoming election. They urged students to register and vote. They passed out fliers on campus.
5. They held a rally the day before the election to focus student attention on voting.
6. On election day, they checked the polls and got lists of people who hadn't voted. They called and encouraged them to vote.
7. The outcome: The voting rate at Stanford was higher than normal. About 65% of the students who said they'd vote did, and many people thanked the students for clearing up the issues.

RESOURCES
- **The United States Student Association,** 1012 14th St. NW, Suite 200, Washington, DC 20005. (202) 347-8772. *Works with students on registration drives. Write or call for info.*
- **Public Interest Research Groups (PIRGS).** Contact: MASSPIRG, 29 Temple Place, Boston, MA 02111. (617) 292-4800. *The PIRGS have registered more students than any other organization.*
- **League of Conservation Voters,** 1707 L St. NW, Suite 550, Washington, DC, 20036. (202) 785-VOTE. *A $25 membership includes a free "environmental scorecard." Scorecard alone is $5.*
- **Election Services Division, League of Women Voters Education Fund,** 1730 M St. NW, Washington, DC 20036. (202) 429-1965. *Sponsors registration drives and supplies printed material.*

2. MUG SHOT

At Potsdam College in New York, use of styrofoam on campus was cut by 58% when students sold reusable mugs in the cafeteria.

It's two in the morning; you take one last sip of coffee and make a hook shot into the trash can. One more paper done…and one more cup headed for a landfill.

Sound familiar? It probably does—the average American student goes through an estimated 500 disposable cups every year.

But there's an alternative to "throwing it all away." You can reduce your dependence on disposables by carrying a reusable mug.

DID YOU KNOW

- Polystyrene foam is an environmental menace. It's often made with ozone-depleting chemicals, it's a hazard to wildlife, and manufacturing it creates pollution. It's also completely non-biodegradable—it just won't go away. Five hundred years from now the foam coffee cup you used this morning will still be around.
- Don't be fooled by plastics industry claims that polystyrene foam is "recyclable." As one observer says: "Recycle means it can be made into the same items again….But it can't—it can only be used to produce various kinds of low-grade plastic. This is not recycling; it is secondary re-use."
- Switching to paper isn't necessarily the answer. Paper cups aren't usually recyclable; producing them uses trees and creates pollution.
- If only 15% of U.S. college students used mugs instead of disposable cups every day, we'd eliminate more than a billion cups a year.
- Cutting down on disposables can save money. California State University students calculated that by eliminating polystyrene, their food services could save $36,000 a year. At the University of Vermont, students project that annual savings could total $76,000.

WHAT YOU CAN DO
On Your Own
- **Precycle.** Cut down on disposables; carry your mug everywhere.

- **Carry a spoon, too**...or ask for one instead of a plastic stirrer.

With Other Students

- **Eco-party.** Next time you have a party, make it BYOC—"Bring Your Own Cup."
- **Push for reusables...and recyclables.** Ask your school food service to use washable china tableware instead of disposable containers whenever possible. If your cafeteria uses take-out containers, ask the food service to make sure they're made of recycled and recyclable, non-laminated paper, too.
- **Organize an all-out ban on styrofoam and nonrecyclable paper products.** At the University of Mississippi, students carried out a successful campaign to replace all of their styrofoam products with 100% recyclable, non-laminated paper plates, bowls, and cups.
- **Start a "mug campaign" on campus.** Educate students about disposables and sell reusable cups. Switching to ceramic mugs isn't practical at some schools. Reusable plastic mugs are an alternative.
- **Encourage your food service and local restaurants to offer BYOC discounts.** UCLA food services allow customers who bring their own cups a 20% discount on all beverages. In one year, this program saved the food services thousands of dollars by eliminating over 100,000 disposable cups. At Cornell, 15 restaurants agreed to offer a discount to students with mugs.

CAMPUS SUCCESS STORY
James Madison University, Harrisonburg, Virginia

Background

A group of JMU students decided to celebrate Earth Day 1990 by selling mugs to help reduce the use of disposables on campus. They planned it to raise awareness about their SEAC chapter...and it turned out to be a successful fundraiser as well.

What They Did

1. They contacted a mug wholesaler (Aladdin) and negotiated a price.

2. They presented a detailed plan and budget to their Student Government Association. The SGA agreed to fund the project.

3. They designed their own logo for the cup, incorporating the James Madison University logo.

4. They ordered 1,000 cups. While they waited for delivery, they met with the school's food services and arranged a cafeteria discount for students who used the mugs. They also convinced five local restaurants to offer the same discount.

5. The shipment of 1,000 mugs was sold out in two days (at $2 a mug). There was a bonus with each mug: a plastic shower curtain ring to attach mugs to belts or backpacks. (The shower curtain rings were purchased locally for 5 cents each.)

6. The mugs became so popular and profitable that the school food service joined in and supported the project. It funded the purchase of more mugs and offered warehouse space. The result: On a campus of 10,000 a total of 7,500 mugs were sold.

7. Success: The mug program became part of school policy. The following year, the JMU administration included a mug in each freshman orientation package. In addition, the food services teamed up with student activists to form an environmental issues committee.

Advice from JMU Activists

- "Try to get permission to use the school's tax-exempt status to avoid paying sales tax."
- "Be sure you can use the school's warehouse or a storage space. Otherwise, a member of your group will end up with wall-to-wall cartons of mugs."
- "If you can't get school funding, try local restaurants. They may pay for your project by advertising on the mugs."

RESOURCES

- **National Wildlife Federation's *Cool It! Program*,** 1400 16th St. NW, Washington, DC 20036. (202) 797-5432. *Can provide a free information packet on how to order mugs and sell them on your campus.*

To order mugs:

- **Whirley Industries,** 618 4th Ave., Warren, PA 16365. (800) 825-5575.
- **Aladdin Synergetics,** 545 Mainstream Dr., Suite 200, Nashville, TN 37228. (800) 888-8018, extension 3630.

3. TOO MUCH FOOD

Americans dump the equivalent of more than 21 million shopping bags full of food into landfills every year.

When you were young, did your parents make you eat everything on your plate? Well, now that you're at school, you don't have to. But what happens to those leftovers?

One expert estimates that Americans throw away more than 870,000 pounds of food each day.

Of course, you can't stop everyone in America from wasting food...but you can help change the way the food service at your school deals with its leftovers.

DID YOU KNOW

- Food waste makes up as much as 50% of the garbage generated by an average school cafeteria.

- On a larger scale: About 12% of the average school's garbage is food waste.

- At Yale University, the waste from food preparation alone amounts to 250-300 gallons per week.

- In a University of Arizona Garbage Project study, students found that 15% of edible food is wasted—most often specialty foods, vegetables, cereal, and grain items.

WHAT YOU CAN DO

On Your Own

- **Take only what you need.** According to a Yale University environmental audit, the majority of food waste in the school comes from student trays.

- **Volunteer for a local food bank.** You'll help your community and learn more about important issues like hunger in America.

With Other Students

- **Start a food recovery program.** The Associated Students of UCLA food service operation sponsors a food recovery program that supplies unused food to a local community center. This program feeds hundreds of homeless people and diverts tons of food from landfills. A group of students at Yale works with a local food bank to collect extra food from their dining hall. They have two food pickups a week and supply 15 to 25 pans of food in each pickup.

Note: According to the Frostburg State University Food Service Director, there are some legal issues concerning the shelf life of food, so a waiver has to be signed by food recipients.

- **Set up a "Vacation Food Alert."** Get your food service to give away perishable food at the start of vacations.

- **Donate a meal.** At Birmingham-Southern College, students organized a "meal ticket donation day." They gave away their tickets for one meal. With the money saved, the food service provided a meal for the homeless.

- **Work with the university to set up a system that converts leftover food into pulp.** Rutgers' food waste goes through a pulping system that grinds it and dries it into a pulp. Rutgers then contracts it to a local pig farmer.

CAMPUS SUCCESS STORY
Carnegie-Mellon University, Pittsburgh, Pennsylvania

Background

A group of Carnegie-Mellon students were shocked by the amount of edible food thrown out at their school every day. Their initial concern grew into a well-organized program to give food to the hungry.

What They Did

1. They asked the director of the Marriott Food Service at Carnegie-Mellon if there was a way to donate excess food to a nearby shelter. The director told them to talk with the university attorney about legal issues.

2. The students met with the attorney, who told them that, given existing legislation, the university could not afford the risk of donating food. According to the law, the university would be held responsible for "simple negligence" if an illness was caused by the food.

3. The students didn't give up. They built support for the food donation program through articles in their college newspaper.

4. A Pennsylvania state representative recognized that existing legislation did not adequately protect the food donor. He worked with the Greater Pittsburgh Community Food Bank to modify the laws on food donation. The new law meant that even if someone became ill after eating the donated food, the university would be legally protected because it did not intend to do harm. When the governor signed the bill into law, the university attorney and the Carnegie-Mellon student who initiated the campaign represented the university.

5. As soon as the law changed, dining services agreed to participate. The day after the bill became law, trucks arrived from the Three Rivers Table, a program that began in 1988 to donate food to local shelters. Food service workers report that bundling the food takes a little extra time...but they feel it's worth it. They've been collecting food virtually every day since the program began. In the six months after the program started in August, 1990, Carnegie-Mellon donated over four tons of food.

RESOURCES

• **National Wildlife Federation's *Cool It!*** 1400 16th St. NW Washington DC 20036-2266. (202) 797-5432. *Call or write for a packet on reducing your food service's environmental impact.*

4. EVERYONE INTO THE POOL

To promote ridesharing, students at Tulane University organized a phone directory that lists people interested in carpooling.

According to a recent survey, colleges issue about 20% more parking stickers than the number of parking spaces they have available.

That may explain why you spend so much time desperately cruising the campus lots, searching for that one elusive parking space. And while you're circling, you're not doing the Earth any favors; your car is spewing out gases that contribute to smog, acid rain, and the greenhouse effect.

What's an alternative? A campus carpool program.

DID YOU KNOW
• Most cars on U.S. roads carry only one person. In fact, we have so much extra room in our 140 million cars that everyone in Western Europe could ride with us.

• If each commuting car carried just one more person, we'd save more than 18 million gallons of gas and keep more than 360 million pounds of carbon dioxide out of the atmosphere...*every day.*

• College campuses deal with a huge number of single-driver vehicles. For instance: Minnesota's Lakewood Community College reportedly averages 5,000 cars a day...though there are only 6,000 students.

• Cars on campuses are major polluters. An environmental audit of UCLA showed that the campus is the 10th largest emitter of carbon dioxide—the main greenhouse gas—in the Los Angeles basin, mostly due to pollutants from cars.

WHAT YOU CAN DO

On Your Own

• **Find out how to connect with other commuters.** Many schools encourage carpools; the office of transportation may be able to help direct you to the right sources. Check to see if there's a commuter ride board.

• **Ask around at classes.** Offer to share your car with people who arrive at the same time you do.

With Other Students

• **Create a ridesharing program.** Match riders and drivers who live within a few miles of one another. One easy method: Place a large map on a bulletin board and have students, staff, and faculty pin their names and phone numbers to their neighborhoods on the map. Make lists of the names and phone numbers of people who live near each other; distribute them to students who want them.

• **Organize a ridesharing directory.** Students at the University of Missouri at St. Louis are sent a list of people who live in their zip code who've filled out a form indicating interest in ridesharing.

• **Pick a central location on campus for ridesharing info.** Put a ride board in the student union or in dorms...Or use the phone. Tulane students run a ridesharing hotline from 9 a.m. to 9 p.m.

• **Provide incentives for ridesharing.** Give students an incentive by providing free or reduced-rate parking and prime parking spaces for ridesharers. At Washington University in St. Louis, parking permits are designed to hang on rearview mirrors so they can be transferred from car to car. This system encourages ridesharing by allowing up to five cars to share a single permit.

• **Work with your state and local governments.** If there are too few student commuters on campus to organize carpools, contact your local government or public library for the names of ridesharing networks in your area. Post and distribute that information.

CAMPUS SUCCESS STORY
Ocean County College, Toms River, New Jersey

Background
On the first day of school, a student noticed that two huge parking

lots, holding more than 4,000 cars, were so full that people had parked on the grass as well. She and another student decided to see what they could do to cut the number of cars on campus.

What They Did

1. The students met with the dean of student life and got permission to poll the student body about their commuting habits. They created a questionnaire and distributed it around campus—on buses, at bus stops, classes, the student center, the library, and all student offices. There were 3,562 responses.

2. They sorted responses by towns, routes, times and dates of travel. They found that 86% of the students surveyed were driving alone.

3. They launched a week-long campaign to build support for carpooling.

4. The students arranged a matching service by writing a carpooling application. Their form included space for people to write their name, address, telephone number, class times, arrival times, times rides were needed, whether they wanted to drive or be driven, and directions from their home to campus.

5. The results: The group began to act as a carpooling referral service. They organized students into carloads of three.

6. As a result of the students' efforts, the college carpool surveys were included in the student handbook. The whole project took one month to establish.

RESOURCES

- **National Wildlife Federation's *Cool It!*** 1400 16th St. NW Washington DC 20036-2266. (202) 797-5432. *Write or call for information on ways to promote ridesharing on campus.*

- **Association for Commuter Transportation,** 808 17th St. NW, Room 200, Washington, DC 20006. (202) 659-0602. *Provides fact sheets on commuting.*

- Check with your state or local transportation departments. Sixteen states have ridesharing programs; nine have special commuter lanes. Local programs can often be found in the phone book under "Rideshare."

5. CUT PAPER

A study conducted by students at Rockingham Community College in Wentworth, North Carolina, revealed that 60% of the school's garbage is paper.

Schools are "paper mills." On the average college campus notebook paper, computer paper, scratch paper, newspapers, magazines, journals, old reports, exams, etc. make up 40-50% of the waste stream.

It's an outrageous waste—paper, especially the high-grade white stock that most schools use, is one of the easiest and most valuable materials to recycle.

Why throw it away when you can recycle it?

DID YOU KNOW
- Every ton of recycled paper saves an estimated 17 trees and 3 cubic yards of landfill space.
- A ton of 100% recycled paper requires 60% less energy and 50% less water during production than paper made from virgin stock.
- Recycling benefits add up: In the first seven months of operation, a UCLA recycling program saved an estimated 1,133 trees and 200 cubic yards of landfill space.
- Schools can help create markets for recycled newsprint: The Rutgers University student newspaper, the *Daily Targum*, uses 100% recycled newsprint. They use approximately 200 tons of recycled newsprint yearly—saving an estimated 3,700 trees a year.

WHAT YOU CAN DO
On Your Own
- **Use both sides of the paper**—whether you're making copies, taking notes, etc. If you have unneeded paper that's printed on one side, turn it into scratch paper (local print shops may even turn scrap paper into pads for you).

- **Find out what kind of paper your school recycles and how it should be prepared.** There are many grades of recyclable paper, from green-bar computer paper (high-quality stock that, in Asia, often replaces wood pulp in papermaking) to "mixed paper" (low-quality stock used in asphalt shingles, egg cartons, cereal boxes). Usually it's necessary to separate materials.

With Other Students

- **Organize to recycle class notes.** Because so much paper was thrown away right after exams, students at the University of Wisconsin in Madison set up a special recycling program during that two-week period. They arranged for the school to put two large trash bins in the student union for recycling. The effort was so successful that it's now a university tradition.
- **Organize a telephone book recycling drive.** Students at the University of Illinois at Urbana started collecting telephone books in early December, covering the 55 buildings on campus and all the dorms systematically. They reclaimed 8,000 directories.
- **Write on white.** Recycling centers pay more for white paper because colored paper is more expensive to reprocess. Solution: Use white paper for everything. Try to persuade all campus organizations to print their announcements on white paper only.
- **Set up a pilot program** showing that recycling paper works. Start with a building or two. Use the scaled-down program to work out problems and streamline the collection and transportation process.
- **Start a large-scale paper recycling program.** Use the pilot program as a springboard to get school-wide recycling established.

CAMPUS SUCCESS STORY
University of Chicago, Chicago, Illinois

Background
A group of students at the University of Chicago set up a pilot program to recycle white office paper. They figured that once it was being recycled regularly, they would expand their program to include other materials.

What They Did

1. The students contacted the Illinois Department of Energy and Natural Resources for help in choosing a reputable recycling collection company. Originally the students wanted to work with a community nonprofit recycling center, but discovered the center couldn't handle large quantities of white paper.

2. The program was formally launched; office paper was collected at seven buildings and sold to an outside, for-profit firm.

3. They immediately started getting calls from campus offices that wanted to participate. After six months, students were collecting office paper in 40 buildings. Figures showed they had collected 135 tons of paper. By checking university purchasing records, students found they were recycling 50 to 60% of the office paper on campus.

4. The success of white paper recycling built student support for an expanded recycling program. "UCRecycle" is now one of the most successful college recycling programs in the country, collecting aluminum, glass, newspapers, cardboard, and cans. It is a formal program funded by the administration.

Advice from University of Chicago Activists

- "The key to success was identifying one or two people in a building who were enthusiastic about recycling and would encourage others in the building to participate."
- "If you can find a nonprofit company that handles white paper, that's the cheapest and best way to go. For-profit waste management companies usually charge base fees; they then return to the client a portion of the money received from the sale of recyclable waste to reprocessors, such as paper mills."

RESOURCES

- **Jack DeBell,** University of Colorado, Campus Box # 207, Boulder, CO 80309, (303) 492-8307. *He's the director of CU Recycling.*
- ***The Recycler's Handbook,*** by the EarthWorks Group. Published by EarthWorks Press, 1400 Shattuck Ave., #25, Berkeley, CA 94709. *$5.95 postpaid, or check with your local bookstore.*
- ***Your Office Paper Recycling Guide***, San Francisco Recycling Program, Room 271 City Hall, San Francisco, CA 94102. *An excellent booklet; $5. Make checks out to City and County of S.F.*

6. LIGHT RIGHT

Yale switched from incandescent to fluorescent lighting in its library. A projected result: Savings of $3.5 million over the next decade.

Have you ever walked by the school library late at night, when no one's there, and noticed that all the lights are on? That's not just a waste of electricity; since power plant emissions contribute to problems like the greenhouse effect and acid rain, it also damages the environment.

That's why it's important to use electricity as efficiently as possible. Turning off lights is a part of the solution, but there are many other ways you—and your school—can learn to "light right."

DID YOU KNOW
• A normal incandescent light bulb is extremely inefficient; 90% of the electricity it consumes becomes heat; only 10% is used for light.
• "Long-life" incandescent bulbs aren't better for the environment. They're actually less efficient than regular ones and can easily cost more in extra energy than they save on replacement bulbs.
• A single 100-watt bulb uses the same amount of energy as four 25-watt bulbs, but emits about twice as much light. It also uses less energy than two 60-watt bulbs, but yields the same light.
• Fluorescent bulbs generally use less energy than incandescents. Example: Brown University students examined 2,200 EXIT signs on campus and found that the school could save $65,000 by replacing incandescents with fluorescent or LED fixtures.

THE COMPACT FLUORESCENT
• The newly developed "compact fluorescent" light bulb gives off light that looks like a normal incandescent bulb's, yet lasts 5-10 times longer and uses as little as 1/4 of the energy.
• Substituting a compact fluorescent bulb for an incandescent will keep a half-ton of carbon dioxide (the main greenhouse gas) out of the atmosphere over the life of the bulb.

- Initially, a compact fluorescent bulb costs as much as $15-$20 more than a standard incandescent. But don't be fooled; over the life of the bulb, the compact fluorescent will actually save you money in reduced energy costs.

WHAT YOU CAN DO
On Your Own
- **When you leave a room, turn off the lights.** People commonly think it takes more energy to turn a light back on than it does to leave it on. That's not true.
- **Dust the bulbs.** Dust on a light bulb or dirt on a glass fixture can reduce the light it gives off by up to 35% and make it seem that you need a brighter, higher-wattage light.
- **Use fewer bulbs in multibulb fixtures.** Remember that one strong bulb is more efficient than several weaker ones. Note: For safety's sake, put burned-out bulbs in empty sockets.
- **Next time you buy a bulb, choose a real energy-saver.** Most "energy-saving" incandescent bulbs just put out less light than their regular counterparts. If you really want to save energy with an incandescent bulb, look for krypton-filled, halogen, or infrared reflective-coated bulbs. Better yet, try a compact fluorescent. Note: Check with your local utility to find out if it offers rebates for these bulbs.

With Other Students
- **Start an awareness campaign.** According to a Harvard study: "Stickers on light switches and a concerted public relations campaign can get students, faculty, and staff thinking about turning off unnecessary lights and appliances. Student environmental groups will gladly distribute the stickers."
- **Work with the administration to find places where lighting can be made more efficient.** For example: libraries, exit signs and hallways. To get things rolling, launch a pilot program. A test run in one dorm might indicate how much energy could be saved before the investment is made across the campus.
- **Encourage the administration to save energy by replacing ordinary on-off switches** in dorm rooms with dimmer switches. (Perhaps as older switches break or as rooms are repainted.) Note: You

can't use dimmer switches with compact fluorescents, so it may only be practical for lights not suitable with fluorescents.

CAMPUS SUCCESS STORY
Brown University, Providence, Rhode Island

Background
In 1990, three environmental science students decided to research "efficient lighting alternatives" for their class project. They met with administration representatives and learned that extensive renovations were planned for four freshman dorms. It was a great opportunity for "practical application of our class research projects."

What They Did
1. They met with a representative of the local power company to find out what incentives were available to the university. Their conclusion: Based on plans for the new dorms, Brown qualified for a "conservation rebate program."

2. To find out about existing lighting, they picked one dorm and conducted a "light bulb audit" there. They recorded the type and wattage of bulbs in each room and surveyed students about the quality of lighting in the dorms. For scientific evidence, they used a technical measurement, called a *footcandle reading*.

3. They went to the power company with their findings. Representatives analyzed the reports and suggested that compact fluorescents would improve efficiency while qualifying Brown for rebates.

4. They contacted lighting distributors to find out how much compact fluorescents cost and how long they'd last. With this info, along with the university's electricity bills for the previous two years, they calculated how long it would take to pay the efficiency investment back...and turn it into real savings.

5. They took their report to the dorm renovation architects and worked with them to integrate lighting suggestions into the plans. The result: Architects put the recommended compact fluorescent lighting in all of the dorms being renovated. Because of the switch, Brown qualified for $40,000 in rebates.

RESOURCES

- **Brown is Green,** PO Box 1943, Brown University, Providence, RI 02912, (401) 863-7837. *Write for info.*

- **Tufts CLEAN (Cooperation, Learning, and Environmental Awareness Now!),** Curtiss Hall, 474 Boston Ave., Medford, MA 02155. (617) 381-3486. *An organization committed to finding ways to reduce energy consumption at the university.*

- ***Home Energy* magazine,** 2124 Kittredge St. #95, Berkeley, CA 94704. *The best magazine in America on home energy. Write for subscription info; send $2 for their excellent "Consumer Guide to Energy-Saving Lights."*

- **Rising Sun Enterprises,** PO Box 586, Snowmass, CO 81654. *Mail orders energy-saving lightbulbs, including some compact fluorescents. Send $5 for their consumer guide / catalog.*

- **Contact SEAC** for more information on student programs.

7. TREE'S COMPANY

Students at Mount Union College in Alliance, Ohio planted 500 trees to restore strip-mined property near the campus. They helped the Alliance community begin an extensive urban reforestation effort.

Most environmental books suggest planting trees—and you already know how important trees are; they help prevent erosion, fight the greenhouse effect, support wildlife, etc. Of course, anybody can plant a tree; but it's easier and more fun when you do it with other people. So why not branch out at your school and get growing?

HOW TO SPELL RE-LEAF
- **Dollars & sense:** The American Forestry Association estimates that in one year, an average 50-year-old urban tree provides $73 worth of air conditioning, $75 worth of erosion/storm water control, $75 in wildlife shelter, and $50 in air pollution control.
- Trees absorb carbon dioxide (CO_2), the main greenhouse gas. A healthy tree can remove between 25 and 45 pounds of CO_2 from the air in a year.
- Trees save energy. According to Global ReLeaf, a well-placed shade tree can reduce the cost of air-conditioning—one of the main uses of energy at educational institutions—by 50%. In the winter, trees act as insulators, reducing energy consumption.
- Students at Little Hoop Community College (located on a Sioux reservation in Fort Totten, North Dakota) put a windbreak of native tree species around their school to help reduce heating costs. It was so successful that Little Hoop is now working with the Bureau of Indian Affairs and the Tribal Council to plant shade trees in other areas of the reservation.

WHAT YOU CAN DO
On Your Own
- **Plant a tree.** Contact your biology or environmental sciences department or your state forestry department for info on selecting and planting trees. Think about native varieties; they'll probably

require less care and are more likely to survive.
- **Make it personal.** For example, offset your own paper consumption by planting a tree every time you subscribe to a new magazine or newspaper.

With Other Students
- **Organize a tree-planting event.** Get permission to plant on campus. To get free trees, check with the agriculture department (if you have one) or your state department of conservation. Example: The New York Department of Environmental Conservation provided the students at Mater Dei College with 3,000 seedlings to reforest land near the campus.
- **Raise money for tree planting.** An "adopt-a-tree" program can help pay expenses; ask people and/or businesses to sponsor trees for a fee that approximates the cost of planting and maintaining them. ...Or turn recycling revenue into trees. At Wichita State University in Kansas, students use proceeds from recycling to buy trees.
- **Get the whole campus involved.** Students at the College of Santa Fe in New Mexico initiated Project Phoenix, a tree- and shrubbery-planting project. Since 45 percent of Santa Fe's student body is Native American or Latino, the Native American Unity Club joined the effort. To help raise money for the project, they sold local jewelry, handbags, T-shirts and mugs.
- **Organize a tree maintenance team.** Students at St. Olaf College in Minnesota planted 4,000 trees on campus for Earth Day 1990. The trees receive continual care through the involvement of environmental studies classes. On Earth Day 1991 students mulched the trees as part of the ongoing maintenance program.

CAMPUS SUCCESS STORY
Central Michigan University, Mt. Pleasant, Michigan
Background
In 1990, a group of CMU students started their own "adopt-a-tree" project. It began as a campaign to mark Earth Day 1990 and combat the greenhouse effect—but it developed into a successful fundraiser.

What They Did

1. The students drew up a proposal for a community-supported tree-planting program. They presented it to the university administration and student union, with a request for "seed money." It was approved. With these funds, they bought 1,000 seedlings and some larger trees from a local nursery and aluminum depression tags that would be attached to the trees to identify each adopter. (Cost: about $30 for 1,000 tags.)

2. Working with the grounds maintenance department and a faculty advisor, they persuaded the university to set aside an acre of land to be designated as an experimental wood lot.

3. Three weeks before the trees arrived, they began promoting the program. Ads ran in school and local newspapers and on local radio stations (which gave them free spots). The ads asked people to donate money to cover the cost of planting and maintaining individual trees. In return, each contributor's name would be put on a tag attached to his or her "adopted" tree. Adoption cost: $3 per tree…or two for $5. The adoptions were managed at a booth on campus.

4. All the trees were adopted—either in advance or on planting day. Tree adopters were invited to come plant their own trees. If they were unable to, the students planted their trees instead.

5. The result: About 950 of the trees are still healthy. Money raised from the adoptions pays for each tree's care for three years, until the trees are okay on their own. Students erected a plaque commemorating the event.

Advice from CMU Activists

- "Organizing a tree planting is a great way to get new people active in the environmental movement."
- "Have trees delivered on planting day. If delivery isn't possible, organize a crew to pick the trees up. And make sure the roots stay wet, or they won't survive."

RESOURCES

- **National Wildlife Federation's *Cool It!*,** 1400 16th St. NW, Washington, DC 20036. (202) 797-5435. *Provides how-to booklet on the ways tree planting can save energy on your campus.*

- **Global ReLeaf,** American Forestry Association, P.O. Box 2000, Washington, D.C. 20013. (202) 667-3300. *Global ReLeaf program has set a goal of planting 100 million trees by 1992. Write or call for information on their campaign.*

- **TreePeople,** 12601 Mulholland Dr., Beverly Hills, CA 90210. (818) 753-4600. *Call or write for info, or send $12.95 ($13.79 in California) + $4.00 shipping and handling for their book: "The Simple Act of Planting a Tree."*

- **"Planting Trees."** Pennsylvania Resources Council, P.O. Box 88, Media, PA 19063. *Instruction pamphlet for tree planting. Send a SASE. Free for single copies or $16.00 for 100 copies.*

8. ENVIRONMENTAL INTERNSHIPS

In a recent poll, a quarter of the college students surveyed said they considered protecting the environment "an important life goal."

America's environmental groups are understaffed, underfinanced, and overworked. They depend on student interns for assistance.

Interns do valuable legal and technical research, help get programs started, and even have a hand in creating legislation.

As an intern, you'll gain practical experience, find out more about what environmental fields you're really interested in, and have a better chance to get an environment-related job when you graduate. You might even get college credits for your work.

DID YOU KNOW

- Internship opportunities are available in most environmental fields, for any length of time—a summer, a semester, or even a year.
- Some internships pay a salary. According to the CEIP Fund, the average wage for the 3,400 students and recent graduates it has placed in internships is $370 a week.
- Interns make a difference. In 1991, for example, three legal interns at the Natural Resources Defense Council (NRDC) started a program to identify environmental hazards in low-income communities. It's now a permanent part of NRDC's work.
- Another example: WorldWatch Institute depends on interns to help produce its annual *State of the World* report.
- Students at Sangamon State University in Springfield, Illinois, hold paid internship positions with the Illinois EPA and other organizations as part of their Graduate Public Service Internship.

WHAT YOU CAN DO

On Your Own

- **Explore internship opportunities.** Talk to professors and your career services office. Check job listing sources (see Resources).

- **Check out the CEIP Fund.** It arranges short-term paid positions in environmental fields for college students or recent graduates. CEIP has four branch offices (Boston, Seattle, San Francisco, and Cleveland) that develop and administer internships in 17 states.

- **Create your own job.** Meet with people where you'd like to work and discuss the organization's needs. Based on the information you get, write a proposal that clearly defines a project, what skills you'll bring to it, and a schedule for it.

- **Get credit for your work.** Talk to your faculty advisor, department chair, or career services office to find out if you can receive internship credit through a class or a major. Some schools require official junior status before allowing credit.

With Other Students

- **Organize an environmental internship day.** Invite environmental groups to send a representative to discuss potential positions.

- **Start an environmental internship bank.** Work with your career services department or environmental studies department to establish a central location for environmental internship info.

- **Find an environmental internship on your campus.** A group of summer interns at the University of Vermont studied the feasibility of a campus composting program. Their effort was part of the school's solid waste reduction program.

CAMPUS SUCCESS STORY
Adam Berrey, Macalester College, St. Paul, Minnesota

Background
Adam Berrey was making plans for the spring semester of his sophomore year when he heard about an internship opportunity through a school environmental group.

What He Did
1. He contacted the sponsor of the internship, the Minnesota Public Interest Research Group (MPIRG). He was told that MPIRG would accept 6 interns; they would work for college credit only.

2. He went to the internship office in the Career Development department to register the internship.

3. He asked a political science professor to be his faculty sponsor. Together, they arranged for the internship to fulfill one course credit in political science.

4. Adam, his faculty advisor, and his employer drafted and signed a "learning contract." It outlined the goals of the project, the strategies to achieve those goals, and the criteria for evaluating Adam's work. The contract included a reading list and required that he write a paper based on his experience.

5. He served as a member of a legislative task force responsible for two bills: one requiring the state government to purchase energy-efficient exit sign lights (which was passed in April 1991); and one that focused on reducing solid waste by regulating packaging. When the packaging bill encountered opposition, he spoke at hearings, met with legislators, and planned a press conference. By the time his internship ended, the issue still wasn't resolved, but he had helped play a crucial role in the battle to pass the bill.

Adam's Advice to Other Interns

- "Keep careful track of the contacts you make—they could help you get a job after you graduate."
- "Look for an internship where you can manage your own project and see a finished product when you're done."

RESOURCES

- **The CEIP Fund,** 68 Harrison Ave., Boston, MA 02111. (617) 426-4375. *Wrote* The Complete Guide to Environmental Careers.
- **SEAC,** P.O. Box 1168, Chapel Hill, NC 27514-1168. (919) 967-4600. *Building a database of internships and job opportunities. Write for info (include a SASE).*
- **National Wildlife Federation's *Cool It!*,** 1400 16th St. NW, Washington, DC 20036, (202) 797-5435. *Has an environmental job bank. Call or write for details.*
- ***Earth Work* magazine,** Student Conservation Association, P.O. Box 550, Charlestown, NH 03603. (703) 524-2441. *Lists job opportunities in the environmental field. Single issue: $6, subscription: $29.*

9. THE RIGHT PATH

The University of California at Davis successfully worked with the Davis city government to establish a comprehensive system of bicycle paths that is "a model for other U.S. cities."

Class starts in ten minutes, so you hop in your car and head for school, combing your hair with your left hand, shifting with your right...and eating breakfast at stoplights.

You're down to the wire, but it looks like you'll make it—until you realize there's no place to park.

Late again! And your gas guzzler is polluting the air, too. There's got to be a better way to travel...and there is: your bike.

DID YOU KNOW

• According to Greenpeace Action, "cars are the biggest source of greenhouse gases and the largest single cause of smog" in the U.S.

• The most damaging auto pollution occurs in the first few minutes a car is running, when the engine is cold. So taking your bike on short trips, instead of your car, can eliminate substantial pollution.

• How fast can you pedal? Studies in Germany show that most bicyclers can travel 10-11 miles per hour.

• According to a recent Harvard transportation audit, parking is one of the biggest issues facing many large campus communities. Most campuses devote between 20-25% of the grounds to roads and parking. Some, like Butler University in Indiana, devote as much as 50%.

WHAT YOU CAN DO

On Your Own

• **Ride your bike to school.**

• **Take it with you.** If you live far from school, throw your bike in the car and use it to get around locally.

• **Contact bicycle organizations in your community.** They may already be working on bicycling issues; you can join them. Bicycle stores may know of clubs in your area.

With Other Students

- **Organize a "bicycle user group"** to improve bicycling conditions at school. Lobby for bike lanes, adequate bicycle parking, and safety education programs for bicyclists.

- **Find ways to make bikes available for free...or at low cost.** At Valparaiso University in Indiana, local individuals and businesses donated 40 bikes to a campus "bike-lending program;" the bikes are checked out by students for the day. The University of California at Berkeley has a used bike exchange. In many school communities, local bike shops have agreed to offer student discounts.

- **Get a local bike shop or campus bike club to sponsor a "Bike Hospital Day."** Students can bring in their bikes to get them repaired and running safely at little or no cost.

- **Work with campus security to protect bicycles.** At the Rhode Island School of Design, students worked with the administration to set up a system of bike engraving and registration. They even arranged for the bookstore to carry an anti-theft bike lock with insurance for up to $1,000.

- **Get support from campus groups.** The administration is more likely to act if there's campus-wide support for your projects, so enlist environmental groups, the student council, professors, etc. It's in everyone's interest to improve bicycling conditions.

CAMPUS SUCCESS STORY

Georgia Tech, Atlanta, Georgia

Background

In 1990, four Georgia Tech students attended SEAC's environmental conference at the University of Illinois. Inspired by the elaborate system of bike paths there, they decided to try to make their own campus bicycle-friendly.

- At the time, there was no place for Georgia Tech students to park and lock their bikes; 397 bikes had been stolen in the last four years (about one every four days).
- Because there were no bike paths, pedestrians had been injured by bikes, and bicyclists had been hit by cars on campus.

What They Did

1. They took photographs and gathered bike-related information at the University of Illinois. They also used a bicycling proposal written by students at the University of Georgia at Athens.

2. At Georgia Tech, they formed a coalition of student organizations to support improved bicycling conditions.

3. They developed a survey and circulated it to bicyclists. The results showed that many cyclists came from off-campus; their main destinations were the student center and classroom buildings; they rode in cold as well as warm weather; and there was unanimous support for decentralized parking.

4. The Institute of Transportation Engineers (civil engineering grad students who were part of the coalition) did a "traffic evaluation," choosing the best routes for bike paths.

5. Some members of the bike coalition had previously worked on recycling and tree-planting programs with a woman at the Office of Facilities. They contacted her again and presented their suggestions for bike parking and paths.

6. She liked their ideas and decided to incorporate them into major campus improvements that Georgia Tech was making for the 1996 Olympics. The first bike racks were scheduled to be installed by the end of 1991.

Advice from Georgia Tech Activists

• "Stay in touch with the people in charge. Keep abreast of the project's progress, and let them know you're still involved."

• "Personal contacts are critical. A good working relationship with a specific individual will often make all the difference in getting administration support."

• "Continuity is important. Make sure there are undergraduates involved to keep the project alive after you've graduated."

RESOURCES

• **SEAC at University of Delaware,** 301 Student Center, Newark, DE 19716. (302) 737-6476. *Can provide sample proposal on how to get bicycle paths, racks, etc. on your campus.*

- **Bicycle Federation of America**, 1818 R Street NW, Washington, DC 20009. *Offers fact sheets and info on how to get more active on a local level. Send a SASE.*

- **League of American Wheelmen**, 6707 Whitestone Rd., Suite 209, Baltimore, MD 21207. (301) 944-3399. *The oldest pro-bike group in America. Write for info.*

- **Pro Bike Directory,** Bicycle Institute of America, 1818 R Street NW, Washington, DC 20009. *Contains program information on individuals, organizations and governmental agencies on a city, state, regional and national level.*

- **Bicycle Legislation in U.S. Cities.** Local Solutions to Global Pollution. 2121 Bonar St., Studio A, Berkeley, CA 94702. (415) 540-8843. *Offers a packet of various cities' bicycle programs and policies for $6.00.*

10. SCALE IT DOWN

Converting to microscale saves a chemistry department an estimated 90% on its supplies and disposal costs.

There's something going on in the chemistry lab you should know about—they're producing hazardous waste.

Even if you're an English major and never set foot in the chem lab, you'll be affected by the pollution it creates.

The solution: Encourage your school to switch to microscale chemistry—a technique that uses miniature equipment and fractional amounts of chemicals to conduct the same experiments as conventional-scale chemistry. This improves the quality of the environment without affecting the quality of education.

DID YOU KNOW

• Almost every college in the country has an introductory organic chemistry course. Each course conducts an average of 25 experiments per year. That's well over 75,000 experiments nationwide in an academic year. Many require hazardous chemicals.

• Chemistry departments generate large amounts of hazardous waste. It is estimated, for example, that a single introductory chemistry course at the University of Arizona generated 4,000 gallons of hazardous waste. The combined cost of purchase and disposal was $13,000.

• Economic reasons for switching to microscale are compelling. Chemical costs have risen, on average, 150% over the past six years. And today it costs between four to six times more to dispose of laboratory chemicals than to buy them. At the Georgia Institute of Technology, costs for disposal of hazardous waste increased from approximately $5,000 in 1982 to over $50,000 in 1989.

• Microscale is gaining academic acceptance. Today about 15% of all colleges in the country have made the switch.

• Four of the major college textbook publishers are producing lab manuals for microscale experiments.

WHAT YOU CAN DO

On Your Own

- **Write a paper** for your chemistry or other science course on the costs and benefits of microscale. Your best source of information: the school's environmental health and safety office or its equivalent. It should have data on the quantity and types of hazardous waste collected and the cost of disposal.
- **Talk to professors and students** in departments that handle hazardous substances about their procedures for waste disposal.
- **Write a letter** to the editors of school and local newspapers outlining the benefits of microscale.

With Other Students

- **Do a cost analysis.** Find out how much your school is currently spending on buying and disposing of chemicals; show the economic benefits of switching to microscale. Another school that has already switched to microscale may be able supply you with figures that show what it has saved in supplies and disposal costs.
- **Set up a campaign** to get your school to switch to microscale. This could include petitions, starting a letter-writing campaign and enlisting community support.
- **Work directly with the chairperson of your chemistry department.** He or she will ultimately be the one to implement the school's conversion to microscale.
- **Cultivate the support of other key faculty.** At the University of California at Berkeley, students in the California Public Interest Reseach Group (CalPIRG) chapter organized student/faculty meetings in an attempt to get the chemistry department to adopt microscale, and used a program initiated by Bowdoin College in Brunswick, Maine, as a model.

CAMPUS SUCCESS STORY
University of Arizona, Tucson, Arizona

Background

In 1990, a group of University of Arizona students conducted an environmental audit of their campus. They found out that their school produced more hazardous waste than any other

Arizona state agency. Their solution: Reduce waste by switching the school's organic chemistry courses to microscale.

What They Did

1. They contacted the U of A chemical supply monitoring system to find out what chemicals the university purchased each semester, and in what quantities.

2. They narrowed their focus to one course: Intro Chem 104—the largest generator of hazardous waste on campus. They used price lists from the school's supplier to calculate the cost of chemicals per student, as well as the total costs for the course.

3. Using information supplied by the Risk Management department, they calculated how much it already cost to dispose of chemicals used in the course. They compared these figures to projected disposal costs if microscale were used. The results: conventional, $13,391 per year; microscale, $267 per year.

4. They worked with faculty to design the microscale lab kits and estimate prices (the equipment is different). The cost for investing in new equipment was $40,000 per 100 students.

5. With the savings of $13,000 per year, they calculated a 3-year payback. Result of the students' work: The university is now considering making the switch to microscale.

Advice from University of Arizona Activists

- "Although your arguments may be compelling, don't expect immediate success. Up-front equipment costs may be too high for some school budgets. Your best approach is to demonstrate the offset costs of reduced chemical purchasing and disposal."

RESOURCES

- **Prof. Dana Mayo,** Dept. of Chemistry, Bowdoin College, Brunswick, ME 04011, (203) 725-3000. *Their chemistry department founded Microscale and provides a newsletter on converting your school's chem department to microscale.*

- **Prof. Kenneth Williamson,** Department of Chemistry, Mount Holyoke College, South Hadley, MA 01075, (413) 538-2349 *Can supply info about textbooks. Teaches microscale to other teachers. Provides textbooks and training for instructors.*

11. DON'T CAN IT

In a 1990 University of Puget Sound survey, 99% of students polled said they would separate and recycle aluminum and glass if a campus recycling program were instituted.

Some students have a drinking problem—they don't know what to do with cans and bottles when they're empty.

Should the containers be recycled...or just thrown away? The choice might seem inconsequential—until you consider the billions of beverage containers Americans dump into landfills every year. Mining, manufacturing, and disposing of them has an enormous impact on the environment.

Another consideration: glass and aluminum come with a lifetime guarantee—they never wear out; the same material can be recycled forever.

DID YOU KNOW

• The energy you save by recycling one aluminum can will operate a TV for 3 hours. The energy saved from recycling one glass bottle could light a 60-watt bulb for four hours.

• Making aluminum with recycled material cuts related pollution by 95%.

• Recycling beverage containers can help pay for processing other recyclables that aren't as marketable. In one month, students at Brandeis University in Waltham, Massachusetts, recycled 5,600 cans, 2,200 glass bottles, and 500 plastic soda bottles. The income they generated paid for recycling five tons of newspaper.

• Students at the University of Alaska at Nome collected aluminum cans and persuaded an Alaskan airline to ship the cans 500 miles—free of charge—to an Anchorage recycling company, which paid the Nome group for each can. They used the proceeds to buy recycling bins for local restaurants.

WHAT YOU CAN DO
On Your Own

• **Recycle your cans and bottles.** Find out where to recycle them on campus.

- **Taking cans and bottles to a local recycling center?** Make the most of your trip; take your friends' beverage containers, too.
- **Send a "Can to Congress."** Join the ongoing campaign; attach a letter to an empty aluminum can and send it to your senators and congressional representative. Ask them to support a national bottle bill that would establish a deposit on beverage containers.
- **If it's possible, buy beverages in refillable bottles.** They can be reused up to 7 times before they need to be recycled.

With Other Students

- **Do a recycling audit.** Find out which areas have the highest concentration of cans and bottles. It can boost your recycling rate.
- **Set up a pilot program.** Prove recycling works on a small scale first. Students at Roger Williams College in Bristol, Rhode Island initiated a pilot recycling program in the science building; the university is now taking over the program. Students at Emory University in Atlanta made $300 recycling aluminum to demonstrate to the university that a new recycling program was feasible.
- **Involve fraternities and sororities.** After Birmingham Southern students identified fraternities as an important source of cans, they created a recycling competition for sororities and fraternities. To encourage participation, the university gives half the money earned by recyling the cans back to the houses. The result: Since the Greek system began recycling, aluminum can recycling has tripled.
- At the University of Florida, students held a recycling competition during Greek Week, with a trophy for the winning house.
- **Work with your environmental studies department.** Students at Texas Southern University in Houston initiated a campus-wide aluminum can recycling project. Led by the environmental health majors, this project was adopted as part of their curriculum.
- **Adopt-a-bin.** Members of an environmental group at the University of North Carolina placed distinctive blue bins for recycling aluminum in dorms, classrooms and the student center. Each member took responsibility for a specific bin to ensure that it was accessible and emptied. Recently an aluminum can crushing contest earned the group over $1,200.

CAMPUS SUCCESS STORY
California Polytechnic University, San Luis Obispo, California

Background

In 1988, a group of students at Cal Poly established an aluminum can recycling program.

What They Did

1. They met with university officials and received permission to conduct a pilot program. They contacted a nonprofit recycling center called Eco-Flow, which provided thirteen 55-gallon drums and paint to identify them as recycling bins.

2. The students collected aluminum cans during a two-day campus open house called "Poly Royal." They collected 179 pounds, which they were able to sell at 30¢ a pound. After the event they asked the administration for permission to expand their program.

3. The students wanted to avoid using fossil fuels when they collected recyclables, so they asked community groups to donate bicycles. They fixed up the bikes with proceeds from the Poly Royal recycling drive and designed bike trailers to haul the cans.

4. In 1990, they received a $28,000 grant from the California Department of Conservation to buy 73 barrels and expand the program again. The result: The university institutionalized campus recycling and began collecting at 59 locations.

5. In its first full year, the recycling program made $2,400. The students reinvested the money in T-shirts for student collectors, publicity, and equipment (e.g., spare parts for the bikes). Now they operate a campus hotline to answer recycling questions.

RESOURCES

• **The National Container Recycling Coalition,** 712 G. St. S.E., Washington, DC 20003. (202) 543-9449

• **The Container Recycling Institute,** 710 G St. S.E., Washington, DC 20003. (202) 543-9449. *Has info on the best recycling methods for aluminum and glass.*

12. HOLD AN ECO-LYMPICS

At Mt. Holyoke College, an energy-saving competition between dorms was called "Make It wEarth It." Energy consumption was reduced by 43,246 kilowatt hours in 3 months—a savings of $4,500.

If you want to play a game of basketball...or poker...or Frisbee football, it's easy to get people involved. They love to compete. But it's a lot harder to get people involved in energy-saving projects. The whole idea sounds like too much work—and they think it's bound to be boring.

Here's a solution: Turn energy-saving activities into a friendly competition. Organize an eco-lympics.

DID YOU KNOW
- Campuses use a lot of energy. For example: San Jose State University, with 21,000 students, has an annual energy bill of almost $4 million—despite the fact that it's in a warm climate.
- In cold climates, energy use is disproportionately higher. The University of Illinois at Urbana has twice the student population of San Jose State, but four times the annual energy bill.
- Eco-lympics save energy. At Colgate University, a competition called "Amps and Lamps" pitted residence halls against each other to cut electric bills. The resulting savings: 200,000 kilowatt hours in just six months.

WHAT YOU CAN DO
On Your Own
- **Be an eco-athlete;** reduce your own energy consumption.

With Other Students
- **Get your school buildings ready for an eco-lympics.** Find out if

they're individually metered. If they aren't, work with your utility company and physical plant operations staff to make the switch. At the University of New Mexico the president supported the idea of measuring conservation efforts—so he arranged to fund the switch to individual metering. Your local utility company might do this installation for free.

- **Hold an energy competition.** At the University of Illinois students held the competition in their sorority and fraternity houses. The SEAC group assigned an energy coordinator in each house to help manage the contest. The Illinois Power Company was so helpful they even offered ten energy audits of the houses for free (normally offered at significant cost) and donated a $200 prize for the winner of the three-month eco-lympics.

CAMPUS SUCCESS STORY
Harvard University, Cambridge, Massachusetts

Background
A group of Harvard students performed a campus environmental audit in 1990 and discovered that dormitory energy use had dramatically increased in the past five years. Many Harvard students have strong allegiance to their residential houses and enjoy a little healthy competition, so they designed an environmental competition to channel this spirit to reduce energy, water, and solid waste. They called it Eco-lympics.

What They Did
1. They contacted the Office of Physical Resources and Operations, which had previously worked with students on an environmental audit. They presented a plan for an Eco-lympics competition that would use the natural division of the 12 dormitories to create the teams.

2. Meter readers. Working with the local power company, they determined that the dorms were metered individually (on many campuses all the dorms are on one meter).

3. Competition. The dorms competed against each other, based on a comparison of their use of energy and other resources in the previous year.

4. Standards. With the help of campus mathematicians, they created a standard consumption figure for each month. This meant considering any factors that might account for discrepancies between the two years. The main issues: the number of students in the dorm and the number of hot and cold days in the month. They called the National Weather Service for assistance.

5. Media campaign. They began the Eco-lympics by publicizing the effort in campus newspapers and on the campus radio station. They also postered the campus and put announcements in house bulletins. The publicity connected the school's dramatic increase in energy costs with the rising cost of tuition.

6. Adminstration support. The dean sent a letter to dorms announcing the competition and outlining the rules. Included with the letter was a phone number students could call with questions or comments.

7. Prizes. After the first month, prizes were awarded to housemasters by a student in a gorilla suit. The winners' names were publicized in campus media, and the dorms with the biggest savings received a Ben and Jerry's ice cream study break. The dorms recorded a savings of 19% on their gas and oil bills.

RESOURCES

- **Mike Lichten, Director of Physical Operations,** Faculty of Arts and Sciences, Harvard University, 1746 Cambridge St., Cambridge, MA 02138. *Can offer advice on how administrators can help students implement energy and resource conservation measures.*
- **National Wildlife Federation's *Cool It!*,** 1400 16th St. NW, Washington, DC 20036-2266. (202) 797-5432. *Offers an Eco-lympics packet.*

13. A ROTTIN' THING TO DO

At the University of Wisconsin at Stevens Point, an estimated 220 tons of compostable material are dumped in landfills every year.

Your school officials are probably looking for ways to eliminate waste and cut costs. So it's a good time for them to start composting...and stop paying to have yard waste and food leftovers hauled to overflowing landfills.

Composting is the process of turning organic material into a rich fertilizer (humus) by piling it up, turning it, and allowing it to decompose. In a compost heap, billions of organisms break organic wastes down into forms that can be used by plants. The finished compost adds nutrients to the soil, improves its texture and increases its ability to hold air and water.

Every school ought to have a composting program; you can help get one started.

DID YOU KNOW

- According to a National Wildlife Federation survey, 25% of the U.S. waste stream is organic and could be composted.
- Composting saves landfill space. Municipal composting, along with curbside recycling, has allowed the university town of Davis, California, to cut its garbage by 50%.
- Composting improves the soil. Students at Southwestern College in Kansas are using composting to restore a five-acre patch of prairie on campus.
- Composting saves money. It costs an estimated $65 a ton to dump solid waste in a landfill; the average cost of municipal composting is only $35 per ton.
- A cost/benefit study at the University of Vermont demonstrated that a composting program could save the school more than $45,000 in just two years.

WHAT YOU CAN DO
On Your Own
• **If you can't have a compost pile in your yard:** You may be able to take yard and kitchen waste to campus or municipal composting sites, farms or community gardens.

• **Advanced composting.** Sort your garbage to separate the organics from the rest; build or buy a small enclosure (your compost bin); learn how to stack and layer the compost; turn it occasionally to accelerate decomposition; allow the air to circulate.

• **Get more info.** Composting is a lot simpler than it might sound, but since we haven't got space to explain it in detail, we recommend that you contact the *Cool-It!* program and ask for their compost packet. (See Resources, or contact the agency that deals with solid waste issues in your state.)

With Other Students
• **Do some research.** Find out what your school does with yard and food waste. Do they have the equipment needed to shred leaves and chip branches? What percentage of their food waste could be composted? Figure out how much money the school will save.

• **Use the research to influence school policy.** Lobby the administration to get a composting program started.

• **Start a pilot program.** Students at Kutztown University in Pennsylvania worked with the cafeteria staff and the Rodale Research Center to design a project using food waste only. They plan to add yard waste once the program is established.

• **Develop a compost program as a school project.** An environmental studies major at the University of California at Santa Cruz started a composting program in his residential dining hall as his senior project.

CAMPUS SUCCESS STORY
Lansing Community College, Lansing, Michigan

Background
A group of LCC students who attended a national student conference at Xavier University in New Orleans was inspired by

discussions with *Cool It!* organizers—and the amount of food being thrown out at the conference—to start a composting program at their own school.

What They Did

1. With the administration's help, the students did a waste audit. They found that 24% of the school's garbage could be composted. They also learned the annual cost of waste removal was $40,000.

2. They worked with statistics and accounting professors to develop a cost savings analysis. They figured out that if they recycled all their recyclable waste, they could save $27,600 in one year.

3. They asked the local government for permission to compost in the city. Officials, worried about pests and odor, said no.

4. They asked the school's president and vice president for permission to compost on LCC's unused land at the airport. They were turned down again.

5. They didn't give up; they arranged a meeting with the city of Lansing's recycling coordinator. He alerted them to a citywide ordinance effective August 1, 1991 banning the disposal of yard waste in landfills. With his support, they came up with a plan for transporting food and yard waste from LCC to a private compost site 10 miles outside of Lansing (which the city already used).

6. They met with the administration again and received a favorable response. As a community college, they wanted to set an example for the community...without spending a lot of money. The students agreed to make composting a volunteer project of their environmental group, provided they could use an LCC truck to take compost materials to the site.

7. They talked to the food service director about putting bins for compostables next to the garbage bins, and developed a schedule for volunteers to make pickups at the dining halls. They also talked to school officials about re-routing yard waste to a central location. In the fall of 1991 students began composting food and yard waste.

RESOURCES
• **National Wildlife Federation's Cool It!**, 1400 16th St. NW, Washington, DC 20036. (202) 797-5435. *Write or call for an informational packet on how you can set up a campus composting program.*

- **Milton DeGraw,** (315) 267-2658. *Director of Food Services at SUNI Potsdam, New York; he's willing to answer students' questions.*
- ***Organic Gardening* magazine,** Rodale Press, E. Minor St., Emmaus, PA 18098-0015, (800) 441-7761. *The monthly source for organic gardeners; $16.97/year.*
- ***Let It Rot: The Gardener's Guide to Composting,*** Story Communications, Schoolhouse Rd., Pownal, VT 05261, (802) 823-5811.
- ***The Rodale Guide to Composting,*** by J. Minnich (Rodale Press, 1979).
- **Contact SEAC** for information on specific programs around the country.

14. SHOW SOME CLASS

Students at the University of Colorado at Boulder set up an Environmental Center that now has over 800 books, 30 periodicals and extensive subject files.

You're interested in taking classes about the environment, but when you look over your school's course schedule, you can only find a few "green" courses…and they don't cover the areas you'd like to study.

So what are you going to do about it?

Why let those in administration be the only ones to decide what courses are available? Join the growing number of students working to make environmental studies a bigger part of their curriculum.

DID YOU KNOW
• There are 125-200 environmental studies programs around the nation.
• A recent American Council on Education poll found that 89% of entering freshmen said the environment was their top social concern.
• A survey by the National Wildlife Federation found that 90% of students believe they and their classmates do not know enough about environmental problems and solutions.
• The same survey found that only 17% of the students felt they learned anything about the environment in class.
• 84% of the students polled said they would take action if they had more information about what to do.

WHAT YOU CAN DO
On Your Own
• **Take an environmental studies course** (if one is available).
• **Create your own project.** Arrange with your faculty advisor to

get course credit for analyzing campus environmental issues and developing more environmentally sensitive policies and programs.

With Other Students

• **Ask a teacher to make environmental action a part of his or her curriculum.** Twenty-five students in a management class at Westbrook College in Portland, Maine, were required to design and manage a campus environmental awareness campaign. Four committees cleaned up a pond, distributed "precycle pads" made from reused computer paper, publicized the campaigns and educated students on environmental issues.

• **Start a student environmental library.** Students at the University of New Hampshire created an Environmental Center with a resource library that serves as a clearinghouse of information on environmental issues. Students at the University of Nebraska at Lincoln worked with their student union to establish a resource center for environmental research.

• **Organize a student-run environmental course.** Engineering students at Tufts University in Medford, Massachusetts, created a class in which they developed pollution-prevention design scenarios for the school. Students at Stockton State College in New Jersey formed an environmental leadership class and received two credits to supervise projects such as: reorganizing their student environmental group, building a nature trail, and developing a campaign to protect the Arctic National Wildlife Refuge.

CAMPUS SUCCESS STORY
University of California, Berkeley, California

Background
A number of students at UC Berkeley wanted to work in the recycling field after graduation, but no university department offered classes in that area. The students set out to design their own course in 1985. They called it "The Joy of Garbage."

What They Did
1. They wrote a plan for a one-semester, three-credit course. Topics included landfills, hazardous waste, radioactive waste, sewage treatment, and more. The course also included independent

projects, field trips, and short internships with local organizations.

2. They worked with DE-Cal (Democratic Education at Cal) and a faculty sponsor to prepare a reading list and syllabus, which was then approved as a full-credit course.

3. They met with community recyclers and local government officials and read recycling journals. They scheduled guest speakers from the community (e.g., a landfill expert). Class members were graded by student coordinators.

4. In order to improve the course, students were asked to write evaluations.

5. "The Joy of Garbage" has continued as a student-run course, with an average of 30 students a year. Each year former students take on the responsibility of coordinating the next year's class.

Advice from UC Berkeley Activists

• "In order to avoid the problem of student turnover, students may want to seek out a faculty member who is willing to teach the class on a more formal, ongoing basis. Note: If no program at your campus allows students to create courses for credit, then work to get such a program started."

RESOURCES

• **Stanford Workshops on Political and Social Issues (SWOPSI),** Innovative Academic Courses, 120 Sweet Hall, Stanford University, CA 94305. (415) 725-0107. *Provides a handbook on organizing student-run environmental courses.*

• **Democratic Education at Cal (DE-Cal),** 320 Eshelman Hall, UC Berkeley, Berkeley, CA 94720. (415) 642-9127. *Helps promote and publicize student-run courses.*

• **Tufts Environmental Literacy Institute,** Tufts University, Medford, MA 02155. (617) 381-3486. *Provides training to faculty and students on creating environmentally literate campuses.*

• **LEAD, U.S.A.,** P.O. Box 275, 22 Spring St., Williamstown, MA 01267. (800) 356-3360. *They do workshops on how students can organize student-run environmental courses.*

15. GREEN YOUR BOOKSTORE

Brown University employees give out over 70,000 plastic bags and 80,000 paper bags per year.

Consumerism certainly isn't the way to save the Earth. The more goods we produce, the more strain we put on the planet. But let's be realistic—we all buy things. And each time we do, we cast a vote for or against the environment.

There are a growing number of environmentally responsible products available—from energy-saving light bulbs to recycled school supplies. Does your campus store stock them?

If not, it's time it did.

GREEN CONSUMING?

• Many wooden pencils are made with rainforest wood...but there are alternatives. For example, the Faber-Castell Company offers "American Naturals," pencils made of "sustained yield wood."

• Signs & Symbols, a California company, sells refillable ballpoint pens made of recycled cardboard.

• A growing number of companies are offering products that help sustain the rainforest by using its renewable resources. For example: Ben & Jerry's Rainforest Crunch, the Body Shop's skin and hair-care products, the San Francisco Hat Company's Panama hats.

• According to a recent survey, 94% of American college students are willing to pay more for environmentally sound products.

• Can college stores provide them? They're trying. The National Association of College Stores (NACS) has formed an environmental task force to determine the best ways that its 3,000 member stores can make a positive environmental impact.

WHAT YOU CAN DO
On Your Own
• **BYOB (Bring Your Own Bag).** Don't use disposable bags; bring

a reusable one or a backpack to carry your purchases.
- **Speak up.** Ask the store manager for environmentally sound products…and buy them when they're available.

With Other Students
- **Put together a catalog of eco-products your bookstore can buy.** Include recycled legal pads, wire-bound notebooks, computer paper, and other office supplies. Also good: environmental books and magazines; string bags; "green" cleaners; unbleached coffee filters, low-flow shower heads, and energy-efficient light bulbs.
- **Set up a bag exchange.** At the Brown University bookstore, reusable cotton string bags are available for customers on an exchange basis. A system of flashing lights reminds them to drop used bags into a bin when they enter the store; they get new bags on the way out.
- **Let people know about it.** If the store manager agrees to buy environmental products, help publicize the effort.
- **Work with your bookstore manager to set up a "Green Section."** Students at Harvard worked with their bookstore to advertise a window sealer that makes student rooms more energy-efficient.
- **Encourage the bookstore to stop filling your bags with "junk ads."** Brown University's bookstore manager stopped stuffing all its bags with newspaper, magazine and credit card advertisements. "We will lose up to $2,000 in revenue by discontinuing the ads," the manager said, "but it's worth it. The longer we wait to reduce waste, the more chance we have of losing customer satisfaction."

CAMPUS SUCCESS STORY
Sangamon State University, Springfield, Illinois

Background
In 1989, members of a student environmental group at Sangamon State noticed there were no recycled paper products and very few environmental books available in the SSU bookstore. In conjunction with a planned celebration of Earth Day 1990, they decided to ask the store manager to remedy the situation.

What They Did

1. The students met with the bookstore manager and told him about the international effort to celebrate Earth Day 1990. They asked him to carry environmental books and recycled paper products, promising that their group would encourage people to buy products from the bookstore if he did.

2. The manager liked the idea and acted on it right away—setting up an environmental book display at the front of the store and promoting recycled spiral notebooks, looseleaf paper and legal pads.

3. The students began their publicity campaign. About a month before Earth Day, they set up a "Countdown to Earth Day" booth next to the bookstore. They put a list of recommended titles on a poster at their booth. They placed bookstore information at busy sections of the campus.

4. The student campaign was so successful that the bookstore set up a permanent environmental section.

5. For Earth Day 1991, the store set up another display, with greeting cards, cloth bags, environmental calendars and T-shirts.

6. The students still promote the bookstore through the Environmental Studies department newsletter. And when an environmental speaker came to campus, the students worked with the store manager to make copies of the speaker's book available. As a result of the students' efforts, the bookstore is known in Springfield as a good source of environmental books.

RESOURCES

- **National Association of College Stores (NACS) Environmental Task Force**, University of Oregon Bookstore, P.O. Box 3176, Eugene OR, 97403. (503) 346-4331. *Call for advice on "greening" campus bookstores. Contact Jim Williams, bookstore General Manager.*
- **Seventh Generation.** Colchester, VT 05446-1672; (800) 456-1177. *Good source for ideas on eco-products your bookstore can stock.*
- **Earth Care Paper Co.**, P.O. Box 7070, Madison, WI 53707. (608) 277-2900. *Sells stationery, etc. printed on recycled paper.*
- **The Official Recycled Products Guide,** P.O Box 577, Ogdensburg, NY 13669. (800) 267-0707. *Costs $105 for an issue; if you don't want to buy it, contact the local library or solid waste office to see if they have a copy.*

16. WATER YOU DOING?

The University of Mississippi uses up to 1.5 million gallons of water per day.

You walk into the bathroom and turn on the shower. While you're waiting for the water to heat up, you realize you've forgotten your towel...so you run back to your room to get it. Meanwhile, the water is still running.

If the water runs for just two minutes before you step into it, you can waste 10-14 gallons. If 10% of America's college students did this just once a year, they'd squander more than 14 million gallons of water.

That's just with a shower. We all use water in lots of other ways—for lawns and gardens, flushing toilets, washing cars. Even if your state isn't one of the many currently experiencing water shortages, we can't afford to waste this precious resource.

DID YOU KNOW
• Americans use an estimated 450 billion gallons of water every day. That's an average of two to four times more water per person than Europeans use.

• Schools use a lot of water. For example: The University of Texas at Austin consumes an estimated 900 million gallons, at a utility cost of over $2 million.

• The University of Illinois is the largest single consumer of water in its district. A Kraft plant nearby uses only 13% of the amount the university does.

• Brown University is the 2nd-largest user of water in Rhode Island.

SHOWERED WITH SAVINGS
• Taking a shower with a standard shower head uses about 5-7 gallons of water per minute. But with a low-flow shower head, you can cut that in half...or more.

- Student surveys have estimated that replacing all of Brown University's shower heads with "low-flow" models would save the university $25,000 a year in water, sewage and energy costs and 11 million gallons of water annually—5% of its water consumption.
- A leaky faucet that fills a coffee cup in 10 minutes will waste an estimated 3,000 gallons of water a year.
- By installing a device called a low-flow aerator on faucets, you can cut the water flow by 25-50% without feeling any difference.
- If you water plants between 9 a.m. and 5 p.m. up to 60% of the water can be lost to evaporation. At the University of California at Santa Cruz, the grounds maintenance switched to watering in the cool hours after 6 p.m. or before 11 a.m.
- Xeriscaping is a new kind of landscaping technique that uses drought-resistant plants and grasses. It can save as much as 54% of the water used in traditional landscaping, keep plants healthier, and improve soil conditions.

WHAT YOU CAN DO

Indoors

- **Do an indoor water audit.** Find out where your school can save water. Contact SEAC for more information.
- **Work with your school** to install water-saving plumbing fixtures in all new buildings and retrofit older buildings.
- **Educate students.** Before installing low-flow shower heads, Brown used a pilot program to determine which model was preferred by students living in dorms.

Outdoors

- **Do an outdoor water audit.** Ask your grounds maintenance crew about water use in irrigation landscaping. Contact SEAC for more information.
- **Change habits.** Ask the grounds maintenance crew to investigate water-saving practices.
- **Suggest "drip irrigation."** According to the Rocky Mountain Institute, a slow drip for an hour, once a week, provides enough water for most trees.
- **Plant native species.** A group of students at the University of Arizona met with faculty and grounds maintenance staff about

water needed for pine trees and other non-native plants. To show that it is easy to plant native species, they planted a baby mesquite tree on the main quad.
- **Promote xeriscaping.**

CAMPUS SUCCESS STORY
Mesa Community College, Mesa, Arizona

Background
Some students and faculty who live in the Sonoran Desert were concerned about the large amounts of water required by the turf grass and non-native trees on their 20,000-student campus. They joined forces to design a campus xeriscape project.

What They Did
1. In the fall of 1989 they submitted a proposal to the administration, outlining the economic and environmental benefits of xeriscaping. The administration offered its full support and helped establish a committee that included representatives from the college, the local water utility, and the city of Mesa.

2. The committee designed a xeriscape landscape to surround a recently constructed science building. The college had planned to install new turf grass around the building, but the administration approved the xeriscape project instead.

3. Students in landscape construction, environmental biology, and natural history of the American Southwest joined with faculty, staff and members of the Environmental Action Committee to begin working on the project.

4. Students picked up donated plants and rocks from local companies. They built raised beds for an herb garden, helped with irrigation, laid granite, and planted trees, shrubs, and cacti. Students in an environmental club help with maintenance, picking up litter, posting educational signs, and announcing volunteer work days.

5. In April 1989, students, faculty and staff officially opened the xersicape garden. The project would normally have cost $150,000 to $200,000. But because of donations and student volunteers, it cost only $12,000. It has already saved an enormous amount in water, fertilizer, and maintenance costs.

RESOURCES
• **Brown Is Green,** P.O. Box 1943, Brown University, Providence, RI 02912. (401) 863-7837. *A campus program that helped the school cut water consumption. Contact James Corless.*

• **South Florida Water Management District,** P.O. Box 24680, W. Palm Beach, FL 33416. (407) 686-8800. *Write for a free color brochure on xeriscaping.*

• **The Texas Water Development Board,** P.O. Box 13231, Capitol Station, Austin, TX 78711. *Write for xeriscape info.*

• **Life Sciences Department,** MESA Community College, 1833 West Southern Ave., Mesa, AZ 85202. *Faculty members will consult with students who want to develop xeriscaping on campus.*

17. GO PUBLIC

Commuting by public transportation takes as little as one-thirtieth of the energy needed to commute by car.

If we want to save the environment, we've got to break our "car addiction." It's pouring pollutants into the atmosphere, depleting the world's oil reserves, contributing to global warming, and causing other ecological damage.

Of course, you can find a lot of reasons to avoid public transportation: It can take a long time, it's hard to keep track of the schedules, it's not private, you always need change, and so on.

But there's at least one compelling reason to "go public"—it's a solution to car pollution.

DID YOU KNOW
• According to the American Public Transit Association, every time you use mass transit instead of your private car, you save an average of 9.1 lbs. of smog-producing hydrocarbons, 62 lbs. of CO_2, and five lbs. of nitrogen oxides, which contribute to acid rain.
• If only 1% of U.S. car owners left their cars home one day a week, it could save 42 million gallons of gas a year and cut pollution.
• A good public transportation system in a college town makes a difference. When the University of Illinois at Urbana improved its transit system, auto use on campus was reduced substantially.
• Campus programs work. When the Lane Transit District implemented a group pass program with the University of Oregon, its ridership more than doubled. When Corvallis Transit tried it with Oregon State, ridership increased 50% system-wide.

WHAT YOU CAN DO
On Your Own

• **Use public transit whenever you can.** If you can't take it all the time, make an effort to take it occasionally—say, one day a week.

- **Pick up a transit route map.** Take a look at the schedule; get a feel for how it fits in with your routine.
- **Get a student bus pass** (a discount pass for students).

With Other Students

- **Set up a central location on campus for bus information.** At the University of Oregon, students arranged for a list of new routes to be distributed in residence halls, the student union, and in the cafeteria. They also placed ads in the campus newspaper and on local radio stations.
- **Work with the school administration to develop new ideas for public transit.** As the result of an environmental audit, University of North Carolina students came up with a new ideas on how to encourage use of mass transit and presented them to the administration.
- **Check out existing bus service.** At a college in New Jersey, students rode buses sponsored by their school to find out why ridership was so low. The answer: The buses didn't travel to the outlying areas of town, where many students lived.
- **Support a student fee.** Propose or petition for a small surcharge to be added to the student bill, so students can have unlimited access to local bus services.

CAMPUS SUCCESS STORY
University of Colorado, Boulder, Colorado

Background
Student executives of the Student Union and Environmental Center noticed problems with Boulder city buses: The routes weren't publicized, they didn't go through campus, and they stopped running too early. As a result, only 2.7% of the university's students used the bus system.

What They Did
1. A student was hired to collect information about mass transit and act as a liaison between the administration, the city, and the

student government. She contacted the University of Oregon and the University of Illinois to find out about their bus pass programs.

2. She met with local government officials and learned of the city's commitment to reduce single-passenger driving in the area by 15% by the year 2000.

3. The Student Union conducted a survey, asking fellow students if they would pay a $10 fee to have unlimited access to local and regional bus service. Of the 400 people questioned, 98% said yes. The findings were presented to city officials, who became more interested in the project.

4. Students formed a committee that included university administration officials, to facilitate the project.

5. The city arranged to bring the directors of the mass transit programs at University of Oregon and University of Illinois to Boulder to advise them on the feasibility of a new program.

6. The university committee and the city met with the local transit authority for the first time to decide what information was needed in order to move forward with the program.

7. They conducted a survey to find out what would motivate students to ride the buses. The results: improved security, lower prices, extended efficient service.

8. In a university-wide referendum, students approved the $10 per student/per semester fee for the bus pass plan. The program provides unlimited local bus use with an expanded schedule to accommodate late-night riders.

RESOURCES

• **University of Colorado (Boulder) Environmental Center.** Campus Box 207, Boulder, CO 80310. (303) 492-8308. *Can provide sample proposals and fact sheets for promoting mass transit on campus.*

• **Contact SEAC** for information on specific programs around the country.

18. PROMOTE ENERGY CONSERVATION

The State University of New York at Buffalo found it could disconnect 50% of the corridor lights in most campus buildings and still have "adequate illumination levels."

There are lots of important reasons to conserve energy: It reduces greenhouse gas emissions, cuts air pollution, protects wildlife and wilderness (by reducing our need to drill and ship oil), limits strip-mining, reduces emissions of sulfur and nitrogen oxides (which contribute to acid rain), and more.

Obviously, saving energy on campus should be a priority. You can help.

DID YOU KNOW

• According to the National Audubon Society, one kilowatt hour of electricity produces 1.5 pounds of carbon dioxide (CO_2)—the main greenhouse gas. At schools, this can add up: An audit at SUNY-Buffalo estimates that the school's use of electricity is responsible for more than 200,000 tons of CO_2—over 7 tons per person—annually.

• Students at Connecticut College found out that reducing room temperature by 1 degree F in every campus building would save 20,000 gallons of fuel oil and $8,000 annually. Temperatures above 68 degrees F require 3% more fuel oil for each degree.

• Princeton University has installed a sophisticated computer control system to monitor energy use in 40 of its buildings.

• Skidmore College has redesigned its heating plant to run entirely on waste crank-case oil.

• Johns Hopkins University heats several of its buildings with waste heat recovered from computer operations.

WHAT YOU CAN DO
On Your Own
• **Turn down the heat.** If your dorm room is too hot or too cold,

call maintenance. Don't avoid the problem by opening your window to send the heat outside.

• **Don't use hot water when warm or cold water will do** (e.g., in washing machines). Heating water is the #2 use of energy in the average American home.

• **Use a clothesline to dry clothes whenever possible.** If you do use a dryer, don't leave it on longer than necessary.

• **Use curtains.** In the winter, they help keep heat inside; in the summer they help keep it out and reduce your air-conditioning needs.

• **Seal your windows.** If you live off-campus and your windows don't seal properly, try an inexpensive sealing putty.

• **Insulate your water heater.** If you live off-campus, install a water-heater blanket.

• **Use a fan instead of air-conditioning wherever possible**—fans are more energy efficient. On cold days, fans can also help move heat around if your room is warmest right next to the register.

• **Turn lights, computers and appliances off.** If they're on when you're not using them, they're wasting energy.

• **Refrigerate right.** Refrigerators use as much as 25% of the energy in an urban apartment. If your refrigerator and freezer are just 10° colder than necessary, your energy consumption will increase up to 25%. Check the temperature: It should be between 38° and 42°; the freezer should be between 0° and 5°. For efficient operation, clean the condenser coils on the back or bottom of your refrigerator at least once a year.

With Other Students

• **Conduct an energy audit.** That's the best way to get a clear picture of what your campus needs to do to save energy. Already, students have conducted energy audits on an estimated 150 college campuses. At Drury College in Springfield, Missouri, the Physics Department has conducted an audit to determine sources of heat loss. Senior environmental policy majors at Howard University met with the campus housing administration to begin conducting an energy audit for buildings on campus...and so on.

• **An audit should determine things like:** Are thermostats at 68°

or lower? Are campus buildings heated all the time, or is the heat turned off nights, weekends, and holidays? Are lights in campus buildings turned off nights, weekends, and holidays? Does the campus use energy-saving light bulbs? What was the campus's total energy bill for the past year? How does your campus compare to other schools? (SEAC can help you find out.)

- **Publicize your efforts.** Use bulletin boards, posters, campus media, etc.
- **Educate others.** At Sonoma State University, the energy center published a flyer on efficient lighting.
- **Look for alternatives.** Find out what sources of alternative energy could be used on your campus. For example: Solar energy might be used to warm buildings or heat water in residence halls.
- **Contact your local utility.** Find out about rebates...and tell the university about them. Many utilities offer incentives for installing energy conservation devices. Contact your local utility to find out what they offer.

CAMPUS SUCCESS STORY
Williams College, Williamstown, Massachusetts

Background
In 1980, Williams students created the Student Energy Conservation Committee. Their long-term goals were making buildings on campus energy-efficient, and teaching other students to be conservation-minded.

What They Did
1. The committee publicized its first scheduled meeting with posters and articles in the campus newspaper. At the meeting, they handed out questionnaires to gauge the interest and experience of students who attended.

2. They held a campus-wide logo contest for their energy conservation campaign. They printed the winning logo, "Think Conservation," on 500 to 1,000 light switch placards and posted them next to light switches around campus.

3. They publicized their efforts. They secured a bulletin board in the student union for posters, meeting announcements, their logo, newspaper articles, etc. They also made radio announcements and put out fact sheets and newsletters.

4. They met with the college president to explain how their efforts could save the college money. They also asked for financial support. In addition, funds were solicited from the physics, geology, chemistry, political science, environmental studies and biology departments—as well as alumni and trustees.

5. They organized activities like tours of a solar house, a local nuclear power plant and the college physical plant. They also organized a campus conservation contest (see Eco-lympics) and a community home weatherization program for low-income senior citizen residences.

6. They published a conservation handbook that was distributed to the entire student body. According to their book, the combined efforts of students and administration led to:

- "An insulation survey of the entire campus, followed by installation of new insulation wherever it was cost-effective."
- "Temperature setbacks in all campus buildings during low-use periods."
- "A campus-wide lighting survey, followed by lowering of light levels in many areas."
- "Caulking, weather-stripping, and curtains being placed around campus."
- "Timer switches being installed in some classrooms."
- "Reduction of hot water storage temperatures."
- "An analysis of energy audit forms that were distributed to and completed by students and faculty."
- "Acquiring funding for a student energy intern position."
- "Hiring students to monitor energy use in non-dorm buildings."

RESOURCES

- **"Recipe for an Effective Campus Energy Conservation Program,"** Union of Concerned Scientists, 26 Church St., Cambridge, MA 02238. (617) 547-5552. *A 17-page booklet about energy conservation efforts at SUNY Buffalo. Provides practical guidelines for establishing similar programs at other universities and large institutions.* $2.00.

- **Contact SEAC** for information on audits and programs around the country.

19. START A RECYCLING PROGRAM

The average student generates an estimated 640 pounds of solid waste in a year; an average of only 5% is recycled.

Let's say there's no recycling program at your school—or if there is one, only a few materials are accepted.

Does that mean you have to take all your recyclables to an off-campus center?

Not necessarily. If you're really committed to recycling, you can work with other students to start a campus-wide recycling program.

DID YOU KNOW

• It's estimated that 90% of all campuses have some recycling in place. But less than half of the programs accept the most common recyclables: newspaper, white and colored paper, aluminum and glass.

• The chief reason for the lack of recycling on campus? According to a UCLA study, there's not enough administrative support.

• Recycling can pay for itself. Rutgers University in New Jersey recycles more than 32% of its waste stream. In one year its program netted over $26,000 and avoided $130,000 in landfill costs.

• At the University of Colorado at Boulder, recycling is paid for in three ways: A small fee is collected from students each semester; the recycling program sells its recyclables; and the university makes a contribution based on its savings in disposal costs.

• Campus recycling programs can provide jobs for students. Birmingham-Southern has 10 part-time staffers, University of Chicago has 6, University of Colorado has 15, and UCLA has 8.

WHAT YOU CAN DO
First Steps

• **Get advice.** Contact schools with successful recycling programs.

• **Look into local and state ordinances.** Are there any that support recycling? You can use them to help establish your program.

- **Check out programs in your community.** You may be able to combine yours with one that's already established.
- **Find out what recyclables can be sold in your area.** If there's no market for recyclables, you end up with waste. Look in the Yellow Pages under "Waste Paper" or "Recycling."

Next Steps

- **Do a garbage audit.** Find out how much your school throws away... and how much of it is recyclable. A garbage audit by students at UCLA showed that over 50% of their campus waste is recyclable. (The majority is high-quality paper.) Students at the University of Chicago found that 75% of their waste is recyclable.
- **Develop an economic argument.** Find out what prices brokers will pay for materials, then look at how much your school already spends on waste disposal. Using this information, demonstrate that your school can save money on disposal and generate income from the sale of recyclables. Be explicit. From the administration's standpoint, this may be the most important argument.
- **Ask for an initial capital investment** to buy necessary materials. At Dickinson College in Carlisle, Pennsylvania, students received a loan from the student senate and then asked the college to match that amount to purchase collection bins.
- **Get supplies.** Your program will require bins for separating and storing materials, a vehicle for transport, a phone number where staff can be reached, advertising supplies (recycled paper, pens, posters), labelling materials to mark the bins (paint, stencils).
- **Get some help.** A successful program must have responsible people to maintain it. Depending on the size of your school, the recycling coordinator may be a full-time employee or a student working part-time. The collection crew can be made up of volunteers who follow a weekly sign-up schedule.
- **If necessary, start small; set up a pilot program.** Sometimes it's better to start on a small scale and work out the bugs before trying to implement a campus-wide program. It's also an effective way to overcome resistance and show the administration that recycling can work. Another alternative: Focus initially on collecting one material. High-grade office paper, especially computer paper, is a good choice since it brings in the most money and is usually plentiful on college campuses (see Cut Paper).

- **Promote your program.** You'd be surprised at what works. Students at the University of Florida set up information tables in front of the Student Union; two students dressed up as "Recycling Gorillas" and reminded anyone who threw their cans in the garbage that it was just as easy to recycle them. At the University of Colorado, recycling posters and door hangers are printed in English, Japanese, Vietnamese and Spanish to ensure that everyone knows how to recycle.

CAMPUS SUCCESS STORY
Texas A&M University, College Station, Texas

Background

A group of students in the Texas Environmental Action Coalition (TEAC) tried to organize a campus-wide recycling program. They were unsuccessful...until they came up with an idea for a pilot recycling program using "dormitory recycling kits."

What They Did

1. They decided what to include in the recycling kits.

2. They put together an information packet detailing the contents of each kit (bins, liners, and a recycling guide explaining how to prepare materials and where to take them in town) and how it would be used.

3. They presented their ideas to the Texas A&M Residence Hall Association. The association agreed to cosponsor the effort, so students were allowed to place recycling bins in the dorms.

4. The students' next task was to finance the operation. They used their PR committee to contact local businesses and solicit support. Businesses were asked to "sponsor" recycling bins. For $25, a local business could have its name (and any other information it requested) stenciled on a bin. The actual cost of the bin and materials was $11—a profit of $14 per bin.

5. In an 8-month span, about 50 businesses participated. With the money raised, the students bought 200 33-gallon Rubbermaid bins from Walmart. They cut holes in the lids and supplied 30 to 50 liners for each dorm. They also included their student-produced recycling guide with each kit.

6. Dorms participated on a voluntary basis—each floor (or wing of a floor) could request a dorm kit. The resident director, the floor resident assistant, and the student in charge of the program signed a contract agreeing to "maintain and recycle materials collected in a clean and orderly way, free from bugs." The students held spot checks and would take phone calls on complaints. The bin would be removed if inspections turned up a problem three times.

7. The results: All the dorms are still participating; no bin has been removed. The students won the Outstanding Recycling Award from the Recycling Coalition of Texas for their dorm kit. The university is now implementing campuswide recycling, and now "dorm kits" are gradually being distributed to local bars, restaurants, and schools—wherever there's a need for recycling.

Advice from Texas A&M Activists

- "Make your collection bins colorful, clean and obvious."
- "Ongoing publicity and education is crucial."

RESOURCES

- **SEAC Campus Environmental Audit Program,** P.O. Box 1168, Chapel Hill, NC 27514-1168. (919) 967-4600. *Provides comprehensive guidelines for setting up campus recycling.*

- **Jack DeBell,** University of Colorado, Campus Box 207, Boulder, CO 80309. (303) 492-8307. *Jack is the school's Director of Recycling and a technical advisor for campus recycling with the National Recycling Coalition. He'll answer your questions on campus waste reduction and recycling.*

- **Texas Environmental Action Coalition (TEAC),** Texas A&M University, Memorial Student Center, Student Finance Center, TEAC #2808, P.O. Box 5668, Aggieland Station, College Station, TX 77844-9081. *Send a SASE for info on starting a dorm recycling program.*

- ***The Recycler's Handbook,*** EarthWorks Press, 1400 Shattuck Ave., #25, Berkeley, CA 94709. *A comprehensive guide to recycling for the beginner. Cost: $5.95 postpaid, or check local bookstores.*

20. BUG OFF

It's estimated that one million people now suffer accidental "severe acute" pesticide poisoning every year.

Y ou've just set up a picnic on the campus lawn when you notice the smell of pesticides.
 That could ruin your meal...so you get up and move.
 But you can't help wondering what poisons you've been exposed to—and whether they really needed to be used in the first place.
 In many cases, they didn't—and you can do something to help your school limit their use.

DID YOU KNOW
• The EPA hasn't finished assessing the health effects of 32 of the 34 most widely used lawn care pesticides. Alarming news: Six of the 32 are targeted for "special review" because they're suspected of causing birth defects, gene mutation, and cancer.

• After broad insecticide applications, a school's lawn may actually be more vulnerable to pest attacks. The reason: Insecticides also kill earthworms (which help keep turf healthy) and beneficial organisms that prey on harmful insects.

• The most effective alternative to extensive use of pesticides is "integrated pest management," or IPM. This includes using biological pest control (e.g., natural predators), pest-resistant plants and limited amounts of pesticides only when absolutely necessary. Note: One expert warns, "Beware of IPM programs that don't really minimize or eliminate pesticide use."

WHAT YOU CAN DO
On Your Own
• **Support alternatives to pesticides.** When you shop, buy organic produce. If your campus food service doesn't carry organic foods, ask the food service manager to start.

• **Write letters to the EPA** (401 M St. SW, Washington, DC, 20460), the Food and Drug Administration (5600 Fishers Ln.,

Rockville, MD 20857) and members of Congress urging them to step up testing produce for pesticides and ban high-risk pesticides.

With Other Students

• **Conduct a pesticide audit.** Find out from the botany, biology, or zoology departments which insect pests and weeds are common to your area and what kind of damage they do. Ask how pests are being controlled at school. (Your grounds maintenance crew will know what chemicals are being used.) Work with the Chemistry department to help identify the chemical components in use and the potential hazards.

• **Educate other students on pesticide use at school.** Write press releases for school or local newspapers, pass out fliers, etc.

• **Encourage the administration to review campus pesticide policies.** They may be receptive...but it may be tough to get their attention: After failing to stop pesticide spraying through administrative channels in 1990, students at Williams College in Williamstown, Massachusetts, pitched tents on their lawn to block herbicide spraying. The next day the administration postponed herbicide application and formed a committee to review its pesticide policies.

• **Help your school initiate an Integrated Pest Management Program.** This may include finding people in your area who practice IPM and learning from them. At Colgate University in Hamilton, New York, students and biology professors identified the site of the worst pest problem and reduced it to spot spraying with only two chemicals. The university cut pesticide spending by 30% and saved $7,000.

CAMPUS SUCCESS STORY

Cornell University, Ithaca, New York

Background

In September 1990, students noticed signs on the campus lawn warning of chemical spraying. A student from a campus environmental group, the Cornell Greens, discovered that Cornell had hired two companies to spray the campus with a variety of turf herbicides, including 2,4-D, a suspected carcinogen; Dicamba, which contains dioxins; and Mecomec, which causes birth defects.

What They Did

1. The Greens asked other students about pesticide use. They found that some had experienced rashes, eye irritation, or headaches—which they attributed to the spraying of chemicals.

2. Did research. Working with the Coalition Against the Misuse of Pesticides, the Greens found out more about the chemicals being sprayed on campus. With this detailed information, they were able to present an educated argument against pesticide use.

3. They circulated a petition protesting the use of herbicides, insecticides and fungicides on campus. They quickly got more than 1,400 signatures. The petitions and a letter were submitted to Cornell's president and several other administration officials.

4. Got publicity. To publicize their campaign, students wrote letters and articles in the Cornell *Daily Sun* and Ithaca newspapers. The Ithaca *Journal* published a long letter from David Nutter, a Cornell ornithologist, expressing anger at the spraying and calling for a citywide ordinance against pesticides. *The Journal* also ran a front-page article quoting Cornell officials and the chemical companies, who said that Cornell complied with all the necessary regulations.

5. Campus workers. The Greens held meetings with Cornell's grounds maintenance workers and asked the president of their union to join them.

6. Met with the faculty. Students arranged meetings with professors in the botany, biology and zoology departments. They expected professors from the school of agriculture to join them, but were turned down. The students were surprised to learn that some ag professors' research is funded by the same chemical companies that produce pesticides.

7. Kept in touch. The Greens sent a letter to everyone who'd signed the petition, letting them know what steps had been taken. They included a letter to parents that could be signed and sent to Cornell's president. They also mailed letters to alumni.

8. Victory: The president promptly agreed to look into the issue and forwarded the petition letter to the associate vice president for facilities and business operations. The vice president agreed to try

several alternatives:
- For lawns: The students put Cornell in touch with a professional organic lawn care expert. Also, Cornell agreed to accept some weeds such as dandelion and clover.
- For trees: Cornell would only spray isolated species of trees that would otherwise die from gypsy moth infestation.
- For sidewalks and gutters: Safer herbicidal soap would be used only where absolutely needed, and there would be an investigation into possible work-study jobs for students to hand-pick weeds.

Advice from Cornell Activists

- "Get as much information as you can, but don't wait to act. You don't have to be an expert to get going on this issue."

- "A petition is a good way to get the word out to a lot of people, build a group of students active in circulating it, and to let those in power know that your side has a lot of supporters."

- "Leave administrators 'room to move'; assume they'll agree with your position, since they're reasonable people and reason is on your side! Be polite but firm, don't take no for an answer, find out what's keeping them from agreeing with you, and deal with it. Get your agreements in writing."

RESOURCES

- **SEAC**, P.O. Box 1168, Chapel Hill, NC 27514-1168. (919) 967-4600. *Ask for their free fact sheet on eliminating pesticides on campus.*

- **Agroecology Program,** Environmental Studies, University of California, Santa Cruz, CA 95064. (408) 459-4140. *Contact: Dr. Stephen Gliessman. A group that fields questions from students interested in setting up agroecology programs on their campuses.*

- **Northwest Coalition for Alternatives to Pesticides (NCAP),** P.O. Box 1363, Eugene, OR 97440, (503) 344-5044. *A clearinghouse for alternatives to pesticides and publisher of the* Journal of Pesticide Reform. *Call or send a SASE for a list of publications.*

- **National Coalition Against the Misuse of Pesticides (NCAMP),** 701 E St. SE, Suite 200, Washington, DC, 20003. (202) 543-5450. *Good source of information on pesticide alternatives. They publish a newsletter called* Pesticides and You. *Send a SASE.*

21. PRECYCLE

The University of Illinois sponsored a shopping trip pitting "recyclers" against "non-recyclers." Each group's purchases were photographed, and the pictures were used to teach students how to shop with waste reduction in mind.

In 1989, the city government of Berkeley, California, initiated a campaign to encourage consumers to buy food packaged in recyclable and recycled materials...or with less packaging to begin with. They called it precycling.

"What we buy has a direct relationship to what we throw away," they explained. "So it's time for a serious examination of what we take home in the first place. Why not reduce waste by *not* buying something? By choosing carefully, we can prevent excessive, unsound materials from getting into our waste stream."

Packaging makes up about a third of what Americans throw away; and it's a luxury we can't afford. So precycling is an important part of your environmental commitment.

DID YOU KNOW

• The average American uses an estimated 190 pounds of plastic each year; about 60 pounds of it is packaging that's discarded as soon as the package is opened.

• According to the Pennsylvania Resources Council: "When we spend our money, we 'vote' for the products that reflect our values. In environmental shopping, every individual's participation does make a difference."

• If 10% of Americans bought products with less plastic packaging just 10% of the time, we'd eliminate some 144 million lbs. of plastic from our landfills, reduce industrial pollution, and send a message to manufacturers that we're serious about alternatives.

WHAT YOU CAN DO
On Your Own

• **Think ahead.** Figure out how you're going to dispose of a product—and its packaging—before you buy it.

- **Think of packaging as part of the product.** You get what you pay for: If the packaging is designed to be thrown away immediately, all you're getting for your money is cleverly designed garbage.
- **Look for containers that can be reused or recycled,** like aluminum and glass, or ones that can be composted, like paper.
- **Buy in bulk whenever you can**—everything from beans to hardware is available without packaging.
- **Avoid items that are made to be thrown away after only a few uses,** like some razors and flashlights. Look for products you can use over and over again—thermos jars, rechargeable batteries, sponges, and so on.

With Other Students

- **Work with your food service.** Suggest that they offer condiments in large dispensers (as an alternative to individual packets), paper goods made of recycled material, etc.
- **Set up a "Minimum Impact Campus" committee.** Students at Connecticut College developed a program using environmental coordinators in dorms to oversee precycling measures like bulk ordering of recycled products (paper towels, napkins, toilet paper), phasing out nonbiodegradable cups, buying phosphate-free detergent, and working with the bookstore to promote recycled products.
- **Use your political clout.** Work with citizen groups such as the Public Interest Research Groups (PIRGs) to support legislation favoring recycling and packaging reduction.

CAMPUS SUCCESS STORY
Dartmouth College, Hanover, New Hampshire

Background
Can you imagine hauling a week's worth of your garbage around campus for seven days? A group of Dartmouth students, faculty members and administrators did that to demonstrate the growing need for precycling and waste reduction.

What They Did
1. Students who'd initiated a campus-wide recycling program were told by the recycling director that it was working...but more had to be done; landfill fees had more than tripled over the last three

years. The students decided to focus their efforts on educating the student body about waste reduction. They wanted to show that solid waste problems affect everyone and that precycling is as important as recycling.

2. They came up with the idea of a "Carry Your Own Garbage Week"—they called it "Trashcapade"—to dramatize the direct effect each student has on the solid waste problem.

3. They recruited volunteers from many campus groups: student government, faculty, minority organizations, fraternities and sororities, campus ministries, athletic teams, etc. A dean and members of the administration also agreed to take part.

4. For one week, 126 volunteer "garbage collectors" carried their trash around campus in two clear plastic bags—one for recyclables, one for nonrecyclables. They hauled the bags with them all day, trying to avoid changing daily behavior so they could accurately gauge the amount of waste they normally produced.

5. At the end of the week, the "collectors" met on the college green for a giant Garbage Weigh-In. The results: They had gathered 550 pounds of trash—350 of which were recyclable.

6. Dartmouth students spread the word about their "Trashcapade" to more than 65 campuses in New England. Students at Trinity College, Wellesley, Harvard, and many others organized their own garbage-carrying weeks.

RESOURCES

- **Dartmouth Recycles,** Recycling Coordinator, McKenzie Hall, Dartmouth College, Hanover, NH 03755. (603) 646-1110. *Pioneered efforts to reduce waste on campus for students, faculty and staff.*

- **"Become an Environmental Shopper."** Pennsylvania Resources Council, 25 West 3rd St., Media, PA 19063. (215) 565-9131. *A handbook and recycled/recyclable products list for $5.*

- **"The Green Consumer Letter,"** Tilden Press, Inc., 1526 Connecticut Ave. N.W., Washington, DC, 20036. (800) 955-GREEN. *A monthly newsletter with the latest info on earth-positive consumer products. Subscriptions: $27 a year.*

- ***Shopping for a Better World,*** Council on Economic Priorities, 30 Irving Place, New York, NY 10003. (800) 822-6435. *A handy guide, $5.95.*

- ***Shopping for a Better Environment,*** *a complete guide to brand name products that won't harm the environment. Check bookstores, or order it for $10.80 from* Meadowbrook Press, 18318 Minnetonka Blvd., Deephaven, Minnesota 55391.

22. BRANCH OUT

In Greenville, South Carolina, students at Furman University worked with local community activists to pass a recycling plan that would take the place of a proposed incinerator.

There are no boundaries to environmental action; you don't have to limit your "green" activities to school. In fact, the more you're willing to get out and work in the community, the more impact your efforts are likely to have.

There's an added bonus, too—community involvement provides a chance to work on a variety of issues with many different types of people. It's a unique opportunity to teach...and to learn.

OFF-CAMPUS
• University of Hawaii students are working with the Rainforest Action Network and other groups to block the drilling for geothermal energy in rainforest areas.

• Students at the University of Colorado worked with the National Toxics Campaign and Citizen Action to encourage Coors, a local brewery, to institute environmentally sound policies.

• Students at Tulane University and the New Orleans SEAC chapter worked with minority groups and workers affected by pollution in the petrochemical industry.

• Boston University students worked with inner-city youth groups to organize environmental education projects at Franklin Park, the largest greenspace in Boston.

• Students from colleges and high schools in Arkansas, Louisiana, and Mississippi protested the proposed incineration of dioxin in Jacksonville, Arkansas.

WHAT YOU CAN DO
On Your Own
• **Find out what community groups work in your area.** Check

the phone book, talk to biology or environmental sciences teachers, call your state department of environmental protection for more information.
- **Join a local, regional or national environmental organization.** Whether groups are organized to save a local stream or save the oceans, they all depend on volunteers to get things done. Contact Public Interest Research Groups, the National Wildlife Federation's *Cool It!* program, the Student Environmental Action Coalition, etc.
- **Do an internship with a community environmental group.** In Berkeley, California, for example, the local Ecology Center welcomes student assistance.

With Other Students
- **Act locally.** *Cool It!* has encouraged college students to help weatherize low-income homes to help the poor better manage rising energy costs. Students do energy audits, install insulation, and caulk windows.
- **Act regionally.** Students at NYU and an estimated 130 campuses in the Northeast have taken action to stop the hydroelectric plant in Quebec. The plant provides electricity to New England; but it's destroying the environment and culture of the Cree Indians.
- **Act nationally.** Key decisions about our environmental future are made in Washington, D.C. Find out from national environmental organizations which issues are most pressing.
- **Act globally.** In 1990, students from eight countries agreed to form an international network called "A SEED"—Action for Solidarity, Equality, Environment and Development—that will use the United Nations Conference on Environment and Development in 1992 to build an international student environmental movement.

CAMPUS SUCCESS STORY
Yavapai Community College, Embry Riddle Aeronautical University, Prescott College in Prescott, Arizona

Background
Yavapai Community College, Embry Riddle Aeronautical University, and Prescott College are all in the same town, but students there had never worked together on environmental issues. The

community-wide celebration of Earth Day 1990 inspired them to establish a Tri-College Coalition.

What They Did

1. Students from all three campuses attended meetings of the Prescott Earth Day Coalition. They decided to organize a community cleanup of Granite Creek, an important riparian area running right though town. It was littered with trash and tires and had become polluted. Since it was a popular picnic and leisure spot, students knew that cleaning it up would attract community support.

2. Students promoted the cleanup in their cafeterias, snack bars, residence halls, and in class announcements. The city government paid for newspaper, radio and television ads.

3. Two hundred people joined in the cleanup, including students, boy scouts, local environmental groups, and the city park commission. The group collected 15 truckloads of garbage.

4. Encouraged by the success of the project, leaders of the three groups formed the Tri-College Coalition. They've continued working together by sending a group of students to a national SEAC conference, organizing to protect Granite Mountain from gold mining, and protecting coyotes.

Advice from Prescott, Arizona, Activists

• "Use the city government's resources. Also, set priorities, and don't try to save the world in one semester."

RESOURCES

• **SEAC,** P.O. Box 1168, Chapel Hill, NC 27514. (919) 967-4600.

• **SEAC International Project: A SEED project (Action for Solidarity, Equality, Environment, Democracy),** Box 686, 532 LaGuardia Place, New York, NY 10012. *An international clearinghouse for young people working on environmental issues.*

• **Public Interest Research Groups (PIRGs).** 215 Pennsylvania Ave. SE, Washington, DC 20003. (202) 546-4707.

• **Citizen Action,** 1300 Connecticut Ave. NW, Suite 401, Washington, DC 20036. (202) 857-5153. *Information on how students can organize grassroots campaigns.*

• **National Toxics Campaign,** 1168 Commonwealth Ave., Boston, MA 02134. (617) 232-0327.

23. BUY RECYCLED

Students at Trinity College in Hartford, Connecticut, organized a successful campaign in 1990 that convinced the college to begin using recycled copier paper and envelopes exclusively.

As WorldWatch says, "There's no cycle in recycle until a throwaway is reused."

Even if you take bottles, cans and paper to a recycling center, you're not *really* recycling unless you use products *made* of recycled materials, too.

By buying recycled goods, students and schools will support the recycling industry—and protect the environment.

DID YOU KNOW
• According to a Stanford University study, only about 50% of all campuses now purchase some type of recycled paper, including bond paper, paper towels and napkins.

• The average university uses over one million sheets of bond and letterhead paper each month. If 1,000 schools bought recycled paper, they'd provide a market for over a billion sheets of quality recycled paper annually.

• Students use over 53 million notebooks every year. If only 10% of the notebooks were made from recycled paper, we'd save an estimated 45,000 trees annually.

• Things are changing. In 1989, there were 30 companies making 170 different recycled products. By 1991, 400 companies were making more than 2,400 different products.

WHAT YOU CAN DO
On Your Own
• **Buy recycled products whenever possible.** Paper isn't the only thing made of recycled materials. Also available: pens, rulers and scissors made partly or entirely of recycled plastic; garbage bags and recycling bins made out of recycled plastic; recycled motor oil for cars or trucks; etc.

• **Be an activist**—get stores to stock recycled products.

With Other Students

- **Check out your school's current buying policy.** See which products can be replaced with recycled items. The best source for info on recycled products: *The Official Recycled Products Guide* (see P. 57). Provide the university's procurement office with the names and phone numbers of recycled products suppliers.

- **Encourage your school newspaper to use recycled paper.** A University of Chicago newspaper uses 7.5 tons of 100% recycled newsprint yearly.

- **Promote the use of recycled paper around your school.** Colorado State University's letterhead is printed on 100% recycled paper. Dartmouth currently uses toilet paper with 80% recycled content and paper towels with 100% recycled fiber. The Environmental Studies Department at Sangamon State University prints its brochure, student handbook and letterhead on recycled paper.

- **Encourage copy centers near your school to stock recycled paper.** Students at Yale University sent a letter to all professors urging them to print course packets on recycled paper; it listed local copy shops that stocked recycled paper.

CAMPUS SUCCESS STORY
Earlham College, Richmond, Indiana

Background

A group of students formed an Earth Day Committee to implement a long-term recycling plan that centered around purchasing recycled products. They started with recycled copier paper.

What They Did

1. They convinced their student government to use recycled paper in its copiers. (They now use 100,000 sheets of recycled copier paper annually.) This showed the administration it could be done.

2. They contacted the Director of Purchasing and explained what resources could be saved by buying 100% post-consumer, unbleached, recycled paper. He agreed to buy recycled paper if students could find it at competitive prices and if a 6-month study showed that recycled stock wouldn't harm the copiers.

3. Comment sheets posted at student copiers showed that students supported using recycled paper, but ran into problems when copying several hundred pages at a time. (The copier heated up and caused a paper jam.) Students considered this a minor inconvenience and agreed to give the copier time to cool during large runs.

4. The students found out how much paper was used in school copiers and how it was purchased. They negotiated with a distributor called Atlantic Recycled Paper and came up with a price only slightly higher than the cost of virgin stock.

5. With the help of the head librarian, the students then convinced the administration to switch its copiers to recycled paper. As a direct result of their effort, students expect the university to adopt a comprehensive recycled paper procurement program.

Advice from Earlham Activists

• Post petitions next to copiers that use recycled paper to show the administration your broad base of support.

• Regional, rather than local, distributors can often offer the most competitive prices. The price differential between recycled paper and virgin stock should be no greater than 10%. Rag bond paper should be equivalent in price.

RESOURCES

• **Dartmouth Recycles,** Recycling Coordinator, McKenzie Hall, Dartmouth College, Hanover, NH 03755. (603) 646-1110. *Can provide info on buying competitively priced recycled, unbleached, post-consumer copier paper, toilet paper and paper towels.*

• **Ray Ching,** Rutgers University Recycling Center, Dudley Road, New Brunswick, NJ 08903. (908) 747-4082. *A technical advisor with the National Recycling Coalition, Ching will advise students on getting campuses to buy recycled products. Call evenings or weekends.*

Recycled paper distributors

• **Earth Care Paper, Inc.,** P.O. Box 3335, Madison, WI 53704, (608) 256-5522.

• **Conservatree Paper Co.,** 10 Lombard St., Suite 250, San Francisco, CA 94111, (415) 433-1000. Orders outside California: (800) 562-9982.

• **Atlantic Recycled Paper Co.,** P.O. Box 11021, Baltimore, MD 21212. (301) 323-2676.

24. WHAT'S FOR DINNER?

The food on the average American's plate has traveled more than 1,300 miles to get there.

Food is an important environmental issue. What you eat, how it's grown, and how it's transported all have an impact on the Earth.

Do you eat meat? A third of the surface of North America is devoted to grazing.

Is your food grown with pesticides? According to the EPA, at least 74 pesticides have been found in the groundwater of 38 states.

Is your food trucked from across the country? Vehicle emissions are responsible for 50% of the smog in metropolitan areas.

We won't tell you what to eat—that's too personal. But we think you should know how your diet affects the environment.

What you do about it is up to you.

TO EAT, OR NOT TO EAT?

• According to *Diet for a New America*: If Americans reduced their meat intake by just 10%, the savings in grains and soybeans could adequately feed 60 million people—the number of people who starve to death worldwide each year.

• According to the *Cool It!* program, students at two Minnesota colleges, Carleton and St. Olaf, found that produce shipped from California to New York consumed 36 times more energy (in fossil fuels) than it provided (in calories).

• Growing grains and vegetables uses less than 5% as much raw materials as meat production does.

• Over 100 active pesticide ingredients are suspected to cause birth defects, cancer, and gene mutation.

WHAT YOU CAN DO
On Your Own
• **Shop at stores in your area that offer locally grown produce.**
• **Buy organic food.** Help reduce pesticide use.

- **Eat low on the food chain.** Even if you're a confirmed meat-eater, you can help protect the environment by cutting down on the amount of meat you eat. It helps if you even cut down a little.

With Other Students
- **Study food service buying patterns.** Students at Hendrix College in Conway, Arkansas, found that only 9% of the school's food came from within Arkansas. Although Arkansas is the largest rice-growing state in the country, the college was buying rice from Mississippi. The school was also importing beef from Texas and Iowa instead of from ranches only a few miles away. As a result of the study, the college increased in-state purchases of beef and rice to 40 percent.
- **Make recommendations to your administration.** Students at St. Olaf and Carleton colleges discovered that only 19 percent of food served on campus was grown in state. Their main food supplier read the students' report and began to purchase from local farms.
- **Meet with the food service director.** Ask if he or she is willing to offer more vegetarian meals and to offer organic produce. Use a petition to show that there's student support for the idea; explain potential economic benefits.
- **Encourage students to use the cafeteria suggestion board.** Lobby for vegetarian meals, organic food, and locally grown food.

CAMPUS SUCCESS STORY
The University of California at Santa Cruz, California

Background
One way to make organic, locally grown food available is to sell it yourself. UC-Santa Cruz students did that by creating a food co-op on campus.

What They Did
1. A group of students in Kresge Residential College, one of 8 colleges on campus, were interested in natural food and cooking for themselves. They met and agreed to form a food co-op to buy their food in bulk.

2. They talked to owners of natural food stores in Santa Cruz to compile lists of distributors of bulk, organic food.

3. They arranged a place for deliveries. The Kresge College office supplied them with a 30' x 100' space with a loading dock. The college charged the students $1 per academic quarter.

4. They contacted the owner of a local health food store, who volunteered to act as a consultant.

5. They bought second-hand equipment from auctions and from stores that were expanding or remodeling. They picked up a cash register, a scale, bulk food bins, a refrigerator case, etc.

6. They buy all organic produce (mushrooms and ginger were the only exceptions) and don't buy products with excessive packaging. They carry no meat products and most produce is from the state. The environmental impact of their merchandise is reevaluated whenever new information is available. They sell items that other students have prepared, including sandwiches, salads, cookies and cakes.

7. The Kresge co-op is a booming success. There are 13 student employees who collectively manage the co-op for $4.40 per hour and get a 25% discount on co-op purchases. Members pay $20/year and receive a 10% discount. Students also receive a 10% discount by volunteering to stock shelves and check in deliveries. Their healthy profit is reinvested in the co-op.

RESOURCES

• **Kresge Co-op**, Box 811, Kresge College, U.C. Santa Cruz, Santa Cruz, CA 95064. (408) 426-1506. *Members can provide assistance for students who want to start a food co-op.*

• **National Wildlife Federation's *Cool It!*,** 1400 16th St. NW, Washington, DC 20036-2266. (202) 797-5432. *Offers a packet of info on how your food choices affect the environment.*

• **Vegetarian Resource Group**, P.O. Box 1463, Baltimore, MD 21203. (301) 366-8343. *Provides meatless recipes for large groups.*

• *Diet for a New America*. *Important book about the environmental impact of America's food choices. Check bookstores, or order from:* EarthSave, 706 Frederick St., Santa Cruz, CA 95062. (408) 423-4069.

• **Democratic Management Services**, 1509 Seabright Ave., Santa Cruz, CA 95062. *Assists students in starting and managing food cooperatives.*

25. BUILD A GREEN HOUSE

Students at Connecticut College started an environmental dorm, known on campus as the Green House. Its purpose is "to serve as an ecological model for both the college and the New London community."

Individuals can make a difference; that was the message of *50 Simple Things You Can Do To Save The Earth*. But if you live in a dormitory you may not have control over your heat, lighting, or water use. You may only have the opportunity to pursue environmental issues as an extracurricular activity.

Isn't there some way you can make environmental action part of everyday college life?

As a matter of fact, there is. It's called an Ecological Living Group, and a growing number of campuses are trying it out. By giving environmental activism a home—literally—on campus, students can develop practical environmental skills and experience.

DID YOU KNOW
• In green houses, students get hands-on experience with solar collectors, recycling systems, composting, food co-ops, gray-water systems, energy conservation, green consuming, etc.

• There are green houses on many campuses, including: Haverford, Bryn Mawr, the University of Oregon, Tufts University, Oberlin College, the University of Pennsylvania, and Reed College.

• Environmental co-ops help students develop commitment. For example: After graduating, alumni of Stanford's Synergy House have gone on to work with EcoNet, the Land Institute, Earth Day 1990, TreePeople, Sierra Club, and the EPA. One former co-op member founded the Working Assets Money Fund.

WHAT YOU CAN DO
On Your Own
• Find out if there's an ecological living group on campus. If

there isn't one, check out the cooperative houses on your campus. Talk to students who live there; explain the idea of an ecological house and see if they're willing to try it.

- **Write to the North American Students of Cooperation (NASCO)** for information on how to start an ecological house on your campus (see Resources).

With Other Students

- **Gather interested people for the project.** Make it a project of your campus environmental group. Give it some time—a group of 15 students at the University of Chicago met for six months before starting their ecological living group.
- **Find some real estate with help from NASCO.** If there are suitable buildings you can rent or lease near campus, then you can follow the steps worked out by the other off-campus student co-ops. NASCO has organized the Campus Co-Op Development Corporation to help students buy their own co-op houses, offering technical assistance and financing from the National Co-op Bank.
- **Get professors involved.** Speak with faculty members about developing an academic program to help legitimize the group and build school support.
- **Offer an environmentally sound meal plan.** By offering a full meal plan, your group can expand to include people living elsewhere as "eating associates," "boarders," or "commuters."

CAMPUS SUCCESS STORY
Duke University, Durham, North Carolina

Background
Students at Duke who'd been part of ecological living groups at other schools decided they did not want to live in dorms, where they only had limited control over their food and living arrangements. They decided to form a green house instead.

What They Did
1. In winter 1989, Duke students created a co-op vegetarian eating club as the first step in building a group for a green house. They found a student house with a seldom-used kitchen and received permission from the facilities department to use it. They registered

as "Meal Plan V," an undergraduate organization. The 25 students cooperatively planned, purchased, and cooked their dinners 3-5 times per week. They used refunds from their dorm meal plans to finance Plan V.

2. The vice president for student affairs supported the eating club. He approached the students and suggested that they expand into a "theme house." Students took the idea one step further—they wrote a proposal for a green house and requested space in a house on campus.

3. They wanted to learn more about ecological living groups, so they successfully petitioned the House Course Committee and took a full-credit, student-run course on the subject.

4. Since there was no house available for their group on campus, the students decided to buy one. They contacted NASCO, which helped them draft a business plan and figure out what they could afford to pay. They were supposed to come up with a percentage of the financing for the green house themselves and get the rest through NASCO after submitting an acceptable business plan. Their offer to buy a house was turned down, but they didn't give up the idea of starting a green house.

5. In fall 1990, the students worked on establishing a green house as an independent project. They recruited new supporters by putting up posters around campus.

6. The president of the university met with them and applauded their efforts to strengthen environmental education. He offered the students a house, if one became available in the future.

7. Finally, they found a house for rent, one block off campus. They moved into it in the summer of 1991.

RESOURCES

- **North American Students of Cooperation (NASCO),** P.O. Box 7715, Ann Arbor, MI 48107, (313) 663-0889. *Services to help students develop housing co-ops.*
- **Synergy House,** c/o 584 Mayfield, Stanford University, Stanford, CA 94305, (415) 723-2300. *Founded the ecological college living group movement in 1971. Publishes a guidebook on starting an ecological living group.*

STAY INVOLVED

Here is a list of some organizations dedicated to protecting the environment. They can supply information and offer support for your efforts.

- **Student Environmental Action Coalition (SEAC)** National Office: P.O. Box 1168, Chapel Hill, NC 27514-1168. (919) 967-4600. Fax: (919) 967-4648.

- **SEAC International (A SEED)**, Box 686, 532 La Guardia Pl., NY, NY 10012. (919) 967-4600.

- **National Wildlife Federation's *Cool It!*** 1400 16th St. NW, Washington, DC 20036-2266. (202) 797-5432. *The group's mission is to motivate college leaders to establish national campus and community models of environmentally sound practices through a culturally inclusive process. Provides resources, organizing tools and regional coordinators.*

NATIONAL STUDENT ORGANIZATIONS

- **American Association of University Students (AAUS)**, 3831 Walnut St., Philadelphia, PA 19104. (215) 387-3100.

- **Campus Outreach and Opportunity League (COOL)**, 386 McNeal Hall, U of MN, St. Paul, MN 55108. (612) 624-3018.

- **National Coalition for Universities in the Public Interest (NCUPI)**, 1801 18th St. NW, Washington, DC 20009. (202) 234-0041.

- **North American Students of Cooperation (NASCO)**, P.O. Box 7715, Ann Arbor, MI 48107. (313) 663-0889.

- **Public Interest Research Groups (PIRG)**, 215 Pennsylvania Ave. SE, Washington, DC 20003. (202) 546-4707.

- **Student Conservation Association (SCA)**, 1800 N. Kent St., Suite 1120, Arlington, VA 22209. (703) 524-2441.

- **United States Student Association (USSA)**, 1012 14th St. NW #207, Washington, DC 20005. (202) 347-GROW.

NATIONAL ENVIRONMENTAL ORGANIZATIONS

• **Citizen's Clearinghouse on Hazardous Waste**, P.O. Box 926, Arlington, VA 22216. (703) 276-7070.

• **Clean Water Action**, 317 Pennsylvania Ave. SE, Washington, DC 20003. (202) 547-1196.

• **Earth Island Institute**, 300 Broadway, Suite 28, San Francisco, CA 94133. (415) 788-3666.

• **Environmental Action**, 1525 New Hampshire NW, Washington, DC 20036. (202) 745-4870.

• **Environmental Defense Fund**, 257 Park Ave. S., NY, NY 10010. (212) 505-2100.

• **Greenpeace USA**, 1436 U. St. NW, Washington, DC 20009. (202) 462-1177.

• **Greens' Committee of Correspondence**, P.O. Box 30208, Kansas City, MO 64112. (816) 931-9366.

• **League of Conservation Voters**, 1150 Connecticut Ave., NW, Suite 201, Washington, DC 20036. (202) 785-8683.

• **National Audubon Society**, 950 3rd Ave., NY, NY 10022. (212) 832-3200.

• **National Coalition Against the Misuse of Pesticides**, 530 7th St. SE, Washington, D.C. 20003.

• **National Toxics Campaign**, 1168 Commonwealth Ave., Boston, MA 02134. (617) 232-0327.

• **Native Americans for a Clean Environment**, P.O. Box 1671, Tehlequah, OK 74465. (918) 458-4322.

• **Natural Resources Defense Council (NRDC)**, 40 W. 20th St., NY, NY 10112. (212) 727-2700.

• **Nature Conservancy**, 1815 N. Lynn St., Arlington, VA 22209. (703) 841-5300.

• **Rainforest Action Network**, 301 Broadway, Suite A, San Francisco, CA 94133. (415) 398-4404.

• **Rocky Mountain Institute**, 1739 Snowmass Creek Rd., Old Snowmass, CO 81654. (302) 927-3128.

• **Save America's Forests**, 1742 18th St. NW, Washington, DC 20009. (202) 667-5150.

- **Sierra Club**, 730 Polk St., San Francisco, CA 94109. (415) 776-2211.
- **TreePeople**, 12601 Mulholland, Beverly Hills, CA 90210. (818) 769-2663.
- **Union of Concerned Scientists**, 26 Church St., Cambridge, MA 02238. (617) 547-5552.
- **Wilderness Society**, 1400 I St. NW, Washington, DC 20005.
- **World Resources Institute**, 1709 NY Ave. NW, 7th Floor, Washington, DC 20006. (202) 638-6300.
- **Worldwatch Institute**, 1776 Massachusetts Ave. NW, Washington, DC 20036. (202) 452-1999.

OTHER RESOURCES

- **EcoNet**, 18 DeBoom St., San Francisco, CA 94107. (415) 442-0220. *A computer-based environmental network, with an eco-bulletin board and an information exchange system designed for college students.*
- **Grassroots Organizing Weekends (GROWS).** *Students in the United States Student Association (USSA) organize weekends to train students to choose good issues and strategies, and build effective campus coalitions. Call or write: USSA, 1012 134th St. NW, #207, Washington, DC 20005. (202) 347-GROW.*
- **Human-i-tees**, 111-115 Tompkins Ave., Pleasantville, NY 10570. (914) 238-6525. *Since May 1990, they have organized environmental T-shirt fund-raisers with over 100 college and high school environmental groups.*
- **SEAC Campus Environmental Audit Project.** *Graduate students from UCLA and SEAC joined forces to develop the Campus Environmental Audit—a blueprint for creating environmental change on campus. To receive a copy of the Audit, or more information on the project, write to the SEAC National Office (listed above).*
- **SEAC Organizing Guide.** *This newly available guide is filled with over 50 pages of info and advice for organizing a SEAC chapter. Call or write the SEAC National Office (listed above).*

JOIN SEAC!

Your membership will help build a strong, unified student environmental movement.

SEAC Membership Benefits

• **Subscription to *Threshold*,** SEAC's national student environmental magazine. Distributed 8 times a year with photos, articles, opinions and ideas for local, regional, national, and international action.

• **Subscription to a SEAC regional newsletter,** published by your regional coordinator several times during the academic year, with updates on local and statewide campaigns.

• **Right to vote and run for office** in SEAC national and regional conferences, and in its elections.

• **Access to SEAC's resources**—fact sheets, campus environmental audit packets, SEAC's organizing guide.

How to Join SEAC

• **For a Student Membership:** Send $15 (or what you can afford) to receive *Threshold* for one year.

• **For a Group Membership:** Send $30 to receive *Threshold* and important mailings for one year. Please include: your group's name, an individual contact's name, your mailing address and phone number, and a list of recent local programs and actions.

• **To Be a Friend of SEAC:** Nonstudents can send $35 to receive *Threshold* and other mailings for one year.

Send to:
**SEAC
P.O. BOX 1168
CHAPEL HILL, NC 27514**